A Thermodynamic /

Tobias Mann

A Thermodynamic Approach to PCR Primer Design

VDM Verlag Dr. Müller

Impressum/Imprint (nur für Deutschland/ only for Germany)
Bibliografische Information der Deutschen Nationalbibliothek: Die Deutsche Nationalbibliothek verzeichnet diese Publikation in der Deutschen Nationalbibliografie; detaillierte bibliografische Daten sind im Internet über http://dnb.d-nb.de abrufbar.

Alle in diesem Buch genannten Marken und Produktnamen unterliegen warenzeichen-, marken- oder patentrechtlichem Schutz bzw. sind Warenzeichen oder eingetragene Warenzeichen der jeweiligen Inhaber. Die Wiedergabe von Marken, Produktnamen, Gebrauchsnamen, Handelsnamen, Warenbezeichnungen u.s.w. in diesem Werk berechtigt auch ohne besondere Kennzeichnung nicht zu der Annahme, dass solche Namen im Sinne der Warenzeichen- und Markenschutzgesetzgebung als frei zu betrachten wären und daher von jedermann benutzt werden dürften.

Coverbild: www.ingimage.com

Verlag: VDM Verlag Dr. Müller GmbH & Co. KG
Dudweiler Landstr. 99, 66123 Saarbrücken, Deutschland
Telefon +49 681 9100-698, Telefax +49 681 9100-988
Email: info@vdm-verlag.de
Zugl.: Seattle, University of Washington, Diss, 2007

Herstellung in Deutschland:
Schaltungsdienst Lange o.H.G., Berlin
Books on Demand GmbH, Norderstedt
Reha GmbH, Saarbrücken
Amazon Distribution GmbH, Leipzig
ISBN: 978-3-8364-6748-3

Imprint (only for USA, GB)
Bibliographic information published by the Deutsche Nationalbibliothek: The Deutsche Nationalbibliothek lists this publication in the Deutsche Nationalbibliografie; detailed bibliographic data are available in the Internet at http://dnb.d-nb.de.

Any brand names and product names mentioned in this book are subject to trademark, brand or patent protection and are trademarks or registered trademarks of their respective holders. The use of brand names, product names, common names, trade names, product descriptions etc. even without a particular marking in this works is in no way to be construed to mean that such names may be regarded as unrestricted in respect of trademark and brand protection legislation and could thus be used by anyone.

Cover image: www.ingimage.com

Publisher: VDM Verlag Dr. Müller GmbH & Co. KG
Dudweiler Landstr. 99, 66123 Saarbrücken, Germany
Phone +49 681 9100-698, Fax +49 681 9100-988
Email: info@vdm-publishing.com

Printed in the U.S.A.
Printed in the U.K. by (see last page)
ISBN: 978-3-8364-6748-3

Copyright © 2011 by the author and VDM Verlag Dr. Müller GmbH & Co. KG
and licensors
All rights reserved. Saarbrücken 2011

A Thermodynamic Approach to PCR Primer Design

Tobias Mann

TABLE OF CONTENTS

	Page
List of Figures	iii
List of Tables	viii

Chapter 1: Introduction	1
1.1 Related Work	3
1.2 Polymerase Extension of Mismatches At the 3' Terminus	10
1.3 DNA Stability	12
1.4 Outline of Thesis	14

Chapter 2: DNA Binding Models	16
2.1 Introduction	16
2.2 Generalized Poland-Scheraga Models	17

Chapter 3: Efficient Identification of DNA Binding Partners in a Sequence Database	27
3.1 Introduction	27
3.2 Algorithms	30
3.3 Methods	40
3.4 Results	41
3.5 Discussion	48
3.6 Future Work	49
3.7 Conclusions	49

Chapter 4: DNA Thermodynamics Applied to PCR Primer Design	50

Chapter 5: Multiple Target Primer Design	78
5.1 Introduction	78

5.2	Methylation Assay	82
5.3	Algorithms	84
5.4	Methods	93
5.5	Results	95
5.6	Discussion	98

Chapter 6:	A Test of Multiple Target Primer Design	99
6.1	Introduction	99
6.2	Primer Design	100
6.3	Methods	100
6.4	Results	103
6.5	Discussion	106

Chapter 7:   Conclusions and Future Work 107

Bibliography 110

Appendix A: Filters 129

LIST OF FIGURES

Figure Number Page

2.1 Nucleic acid binding affinity. The Generalized Poland-Scheraga models of DNA binding express the binding affinity between two molecules as the sum of stabilities of all conformations in which there is at least one base pair (represented with a dot) between the two molecules. (A) The stability of a conformation is decomposed into a product of terms accounting for the stability of unbound ends, helical regions, internal loops (left), and bulge loops (right). (B) The partition function, Z, is the sum of stabilities of conformations with at least one base pair. . . 18

2.2 Two conformation of a DNA dimer. (A) An allowed configuration, in which the sequences are bound. (B) A forbidden conformation with a pseudo-knot. 20

3.1 Overview of filtering algorithm. **(A)** k**-mer filtering.** All k-mers for a specified value of k are input to the mismatch filter, along with a set of pre-chosen similarity thresholds. The four filters eliminate k-mers in turn, producing as output a list of candidate k-mers that could anchor a binding site. We subject all k-mers to two sets of thresholds, producing two sets of candidates binding site anchors. One set yields k-mers that have high thermodynamic affinity to the query, and the other set yields k-mers that have high sequence similarity to the query. **(B) Candidate retrieval and evaluation.** The k-mers that passed the filtering steps in (A) are located in the genome sequence using a precomputed index. We examine only those sites where a candidate k-mer from one group occurs with close proximity to a candidate k-mer from the other group. These candidate binding sites are then tested for binding affinity using the partition function model, and all sequences that bind to the query with greater than a target affinity are reported. 28

3.2 Filtering k-mers. **(A) Decomposition of the Query sequence into k-mers.** The query sequence is decomposed into overlapping k-mers of a specified length. **(B) Computation of the similarity of a k-mer to the query.** Each filter identifies k-mers that could anchor a binding site, taking as input the k-mers derived from the query sequence, a candidate k-mer, and a pre-specified filter threshold. Each filter then reports whether the candidate k-mer had the specified level of similarity to at least one of the k-mers in the query sequence or not. 33

3.3 Search for proximal hits. Our binding site search algorithm finds anchoring k-mers in the search sequence. We use two sets of filter thresholds, and obtain two sets of candidate anchoring k-mers; one set has high similarity to the query, and the other set has high affinity to the query (occurrences of k-mers from the high affinity set are drawn with dashes above the search sequence, and occurrences of k-mers with high similarity are drawn with solid lines below the search sequence). We locate all occurrences of both groups of candidate anchoring k-mers, and further examine only those sites where there is a candidate anchoring k-mer from the high similarity group occurring within a pre-specified distance w from a candidate anchoring k-mer from the high affinity group. 37

3.4 Filter Analysis of a binding site. **(A) Decomposition of binding site.** Each binding site is decomposed into its constituent k-mers. **(B) Ranking of binding site k-mers according to thermodynamic affinity.** The binding site k-mers are ranked according to similarity to the query by F4; F3 is used to break ties. The similarity scores of the top ranked k-mer are added to the set of similarity scores used to determine filter thresholds for the high affinity group of candidate k-mer binding site anchors. **(C) Ranking of binding site k-mers according to sequence similarity.** All k-mers, except the top ranked k-mer in (B) are re-ranked according to sequence similarity to the query by F1; F2 is used to break ties.The similarity scores of the top ranked k-mer are added to the set of similarity scores used to determine filter thresholds for the high similarity group of candidate k-mer binding site anchors. 38

3.5 Example of weak binding site. ΔG is -10.9 kcal/mole 46

3.6 Example of medium site. ΔG is -16 kcal/mole. 47

4.1 Flowchart of the Pythia algorithm. Inputs are the genomic sequence, locus coordinates, and user specified parameters. In step 1, Pythia identifies all primer pairs meeting the user specified requirements and sorts these primer pairs by the sum of the differences between the computed and target primer melting temperatures. In step 2, Pythia computes the thermodynamic quality metric for the top ranked candidate. If this candidate meets a user specified metric threshold, then Pythia proceeds to step 3. If not, the top ranked candidate is removed from the list and Pythia returns to step 2. In step 3, Pythia performs a specificity check. If the primer passes the specificity check, it is given to the user, and the program terminates. If not, the top ranked candidate is removed from the list and Pythia returns to step 2. 54

4.2 Species accounted for in primer feasibility analysis. The solid line is the top strand of the template;the dashed line is the bottom strand of the template; the arrow with the square end is the left primer; the arrow with the round end is the right primer; three dashed lines indicate binding (or folding) via hydrogen bonding. (A) Desired binding interactions. High rates of binding are desired between the primers and the template priming regions. (B) Undesired binding and folding reactions. Primers should not fold, dimerize, or bind to the target outside of the priming regions. 56

4.3 Primer pair design coverages for interspersed repeat regions. Design coverage is defined as the fraction of an interval covered by PCR product sequences. (A) Histogram of coverages for Pythia (mean 80%). (B) Histogram of coverages for P316 (mean 50%). 66

5.1 Algorithm for multiple target primer design. The input are a set of loci, and the output is a set of primers that will prime each of the loci; our objective is to make the primer set as small as possible. The algorithm works in 4 steps. 1: A sequence not targeted by a primer, call the centroid, is chosen. 2: Untargeted loci similar to the centroid are identified. 3: Primers are optimized for binding affinity to each element of the set. 4: The list of targeted elements is updated based on the sequences to which the optimized primers will stably bind. . . 80

5.2 Bisulfite conversion assay. (A) First, genomic DNA is cut with a methylation insensitive restriction enzyme that cuts in a locus of interest. (B) A hairpin linker is then ligated to the cut-site. (C) the DNA is then subjected to bisulfite conversion, which changes unmethylated cytosines to uracil, but leaves methylated cytosines unchanged. These samples are then PCR amplified,cloned, and sequenced to identify methlyated cytosines. 83

5.3 Illustration of our algorithm for multiple target primer design. The input is a set of sequences for which primers must be designed, and the output is a set of primers that amplify multiple elements in the set. (A) We enumerate all ordered pairs of 5-mers and identify all loci in which the first word occurs 5' of the second word. The sequences that a primer must target for each locus and word pair are the reverse complement of the 30 base pair sequence ending at the first word, and the 30 base pair sequence beginning at the second word. Loci are discarded if either candidate primer target contains a CpG dinucleotide. (B) Our primer design algorithm has three parts. First, a locus is chosen at random and designated as the centroid; this is the first element in the set of candidate targets. All other loci are compared to the centroid, and added to the set if there is at most one mismatch between each target terminus and the centroid, and if the thermodynamic similarity between a locus and the centroid, according to our thermodynamic similarity function, is above a threshold. Second, two graphs are formed whose nodes are the left and right targets. Two nodes in a graph are connected by an edge if their thermodynamic similarity is above a threshold. In this step, loci are identified with the property that their left sequences form a clique in the left target graph, and their right sequences form a clique in the right target graph. These cliques are identified using a one-class support vector machine. Finally, primer sequences are optimized to bind to the reverse complements of the sequences in the left clique and to the sequences in the right clique. . 85

5.4 Our thermodynamic definition of similarity is based on the fraction of strands of sequence A (solid line) bound to the reverse complement of sequence B (dashed line) for a system in which the strands were present at equal initial concentrations s_0. If A is thermodynamically similar to B, then at equilibrium the concentration h of dimers will be high, and the concentrations of the single strands $s_0 - h$ will be low. 87

5.5 Our similarity function produces a similarity metric that lies strictly between 0 and 1. Because we use it at temperatures such that most strands will be in the dimer form for a sequence and its reverse complement, the norms of the data under the embedding are all close to 1, and the data can be regarded as located near the surface of a hypersphere. In this context, training the SVM to separate data from the origin is equivalent to finding a small ball that contains most of the data. 90

5.6 Number of primable loci as a function of primer pairs. Dashed line is the coverage assuming one primer pair per locus ($y = x$); Solid line is number of primable loci using multi-target primers. All loci can be primed using 204 primer pairs. 96

5.7 Number of primable cut sites (with the 5' segment treated as distinct from the 3' segment) as a function of primer pairs. Dashed line is the coverage assuming one primer pair per locus ($y = x$); Solid line is number of primable loci using multi-target primers. All 2500 loci can be primed using approximately 1500 primer pairs. 97

6.1 Gel showing results of amplification from purified PCR products. See text for explanation of gel lanes. Lanes marked "L" are 100 base pair ladders. 104

LIST OF TABLES

Table Number · Page

3.1 ΔG PCR: Free energy of binding, in kilocalories per mole, of the sequence to its reverse complement at 55 C in 50 mM NaCl and 2 mM $MgCl_2$; ΔG MA: Free energy of binding, in kilocalories per mole, of the sequence to its reverse complement at 40 C in 1 M NaCl. Energies are computed using the HYBRID software. 40

3.2 Rejection rates for the four filters. The table lists, for each of the query sequences in Table 3.3, the percentage of k-mers rejected by each of the four filters using the high affinity filter thresholds, as well as the total number of k-mers that pass through all four filters. These results are for weak binding sites in standard PCR conditions. 42

3.3 k-mer filtering performance. The table lists, for each of the query sequences in Table 3.3, the total number of k-mers that passed through all four filters using the high affinity thresholds. 43

3.4 Proximity filtering performance. The table lists, for each of the query sequences in Table 3.3, the percentage of sequence locations that were rejected by the proximity filtering step. The final row contains the column average. 44

3.5 Number of candidate sequences examined and accepted by the partition function model of DNA binding, and time for each run. The table lists, for each experiment, the total number of candidate sites produced by the filtering and indexing pipeline, the number of those sites that are considered by HYBRID to be true binding sites, and the total wall clock time required to identify the sites. 45

4.1 Concordance between agarose gel and PCR amplicon melting curve results. For a selected set of PCR primers, we compared the results of melting curve analysis to agarose gel analysis of PCR amplicon. Melting curves were classified as valid or invalid based on melting curve morphology, and gel lanes were classified as clean or not clean at two levels of stringency. In each table entry, the numbers correspond to the number of reactions with the corresponding gel and melting curve label at stringent and permissive levels of gel scoring stringency. . . 61

4.2 Genomic characteristics of selected human genome regions. We compared the ability of Pythia to the ability of the P316 algorithm to tile these regions. We show the location, size, and a brief description of each locus. 62

4.3 Primer design performance for selected human regions. Shown are the number of PCRs and the extrapolated success rates for permissive and stringent criteria. 63

4.4 The number of training points for each acceptability threshold. For each threshold, we show the number of examples used to train the SVM, and the ROC and ROC50 scores. We assessed SVM performance using 5-fold cross validation. 65

4.5 Pythia acceptability assessment of P316 primers. We assessed the ability of the Pythia primer pair quality metric to predict the quality of the P316 primers. Here, we compared the Pythia primer assessment to the results of melting curve analysis. 67

4.6 Primer3 acceptability assessment of Pythia primers. We assessed the ability of the Primer3 scoring function to predict the quality of the Pythia primers. Here we compare the P316 primer assessment to the results of melting curve analysis. 68

4.7 Pythia specificity assessment of P316 primers. We assessed the ability of the Pythia specificity assessment to predict which amplicons would yield non-specific products. Shown is the results of the Pythia specificity assessment of P316 primers compared to the labels of melting curves. 68

4.8 P316 specificity assessment of Pythia primers. We assessed the ability of the P316 specificity assessment to predict which amplicons would yield non-specific products. Shown is the results of the P316 specificity assessment of Pythia primers compared to the labels of melting curves. 69

5.1 Methylation Insensitive Restriction Enzymes. 93

ix

5.2	Mean number of primable sequences for varying joint and centroid similarity. Each image is a heat map, with the joint similarity on the x-axis and the centroid similarity on the y-axis.	95
6.1	Expected products. Column 1 is the primer pair ID. Column 2 is the product ID (3 for each primer pair). Columns 3 and 4 are the free energies of binding between the left primer and the target at 55 C and 45 C, respectively. Columns 5 and 6 are the free energies of binding between the right primer and its target at 55 C and 45 C, respectively. Column 7 is the product length.	101
6.2	Primer sequences used for experiments.	101
6.3	Genomic coordinates of the expected products.	101
6.4	Primer sequences for resequencing analysis.	102
6.5	PCR products recovered.	103
6.6	Primers and templates for gel lanes.	105
7.1	Loop penalty matrix for Alignment Filter.	130

Chapter 1
INTRODUCTION

The polymerase chain reaction (PCR) is a method for making many copies of DNA molecules with a specific sequence in a complex mixture of DNA molecules. The essential insight of PCR is that biological enzymes and substrates can be used in conjunction with careful control of temperature in order to make exponentially many copies of a target molecule with repeated reaction cycles. This capability has many applications; just a few examples are genotyping, preparing a sample for DNA sequencing, and microarray manufacture. PCR quickly became a fundamental tool in molecular biology, so much so that its inventor, Kary Mullis, was awarded the Nobel prize in Chemistry in 1993.

Crucial to the success of a PCR is the design of two short DNA molecules that flank the fragment to be copied; these sequences are called the PCR primers. These primers must satisfy a number of criteria, and the design of these primers has attracted sustained and ongoing research. Early research focused on basic methods for choosing primers, and later research focused on applications of these basic methods for a variety of PCR applications.

A method for choosing primers must balance a number of criteria. First, the primers must bind well enough to the target molecules to enable extension by a DNA polymerase, but not so strongly that they cannot be denatured at temperatures compatible with the DNA polymerase enzyme. Second, the primers must not bind to each other, or fold into stable secondary structures. Third, the primers must also not bind with good affinity to the molecules to be amplified outside of the priming region. Finally, the primers must not cooperate to amplify undesired DNA, although this requirement is more important in some PCR applications than others.

Early researchers who developed the initial PCR primer design methods knew that DNA binding and folding considerations are fundamentally important in primer selection; however, principled methods to compute DNA binding and folding stability have only recently become available [60, 45]. Two crucial developments enabled a principled consideration of DNA binding and folding for PCR primer design. The first was a comprehensive measurement of DNA stability parameters for a variety of sequence motifs, including not only stabilities attributable to hydrogen bonding and base stacking, but also sequence specific mismatch stabilities and stability data on unbound bases in a DNA helix. The second development was a statistical mechanical approach for computing the overall folding stability for one and binding stability for two arbitrary DNA molecules. This comprehensive set of DNA stability data and a principled method to use this data to compute binding stability enabled one of the principle objectives of this work, which is a comprehensive incorporation of DNA binding and folding considerations into PCR primer design.

Early researchers were also aware that primer specificity (the degree to which a primer pair will direct amplification of sequences other than the intended sequence) was a serious concern. However, this concern could not be addressed without knowing the set of DNA fragments in a reaction. Because these other fragments were typically fragments of genomic DNA, this concern could not be comprehensively addressed until genome sequences were determined. Even when genome sequences were available, it was still unclear how to search them for primer binding sites. Because efficient tools were available for searching genomic sequences for homologous sequences to query sequences, these tools were also commonly used for primer binding site search. The use of these tools is reasonable, because sequence similarity is a pre-requisite for binding stability, but these tools also lack sensitivity, because alignments employed for homology differ in fundamental respects from alignments employed for stability computations. An important objective of this work was the development of principled methods for searching a genomic database for binding sites to primer sequences.

This work has two main objectives as described above: a principled method for incorporating DNA binding and folding stabilities into PCR primer design, and a sensitive method for identifying DNA binding sites in genomic DNA. The motivation for these objectives is to improve the reliability and specificity of PCR. The approaches described in this thesis were experimentally tested and shown to work well. Furthermore, these approaches can be readily extended for more challenging primer design tasks in a way that more heuristic methods cannot; one example of this is given in a chapter on multiple-target primer design, in which the goal is to design primers that amplify more than one target.

1.1 Related Work

Research in primer design started shortly after the advent of PCRs with polymerases able to maintain activity after exposure to high temperature [170]. Before the use of thermostable polymerases, polymerase enzyme had to be added at each cycle of a PCR. Upon the discovery of thermostable enzymes [170], the PCR protocols became greatly simplified, and the use of PCR greatly increased. This increasing use of PCR, in a wide variety of contexts, motivated research into methods to reliably choose effective PCR primers.

1.1.1 Basic Approaches to Primer Design

The early papers in primer design started from the concept of a score, used to quantify the quality of a primer. As stated by Rychlik and Rhoads, the quality of a primer was essentially "the ability of the primer to form a stable duplex with the specific site on the target DNA, and no duplex formation with another primer molecule or no hybridization at any other target site" [167]. These quality scores take into account a number of criteria that a good primer pair should satisfy, and typically include a

1.1.2 Applications of the Scoring Function Approach

After the scoring functions for primer pairs were developed, researchers then applied these functions to the design of primers for specialized applications. Ping et al. [100] applied the scoring function approach, in combination with sequence quality data from DNA sequencers, to automatically design PCR primers to fill gaps during genomic sequencing. Varotto et al. [200] developed a method to design primers to amplify segments of genes that do not overlap introns on a genomic scale; this method was notable for incorporating thermodynamic models of DNA folding into primer evaluation.

Zakour et al. [221] developed a method to design primers to tile bacterial genomes in 10 kilobase fragments. They used standard primer scoring methods; however, in contrast to available methods, their approach was able to take a genome-length sequence and tile it with long amplicons. This work was one of the first to focus on primer design for high throughput applications, in which all the primers needed to have homogeneous thermodynamic properties so all PCRs could be run under the same reaction conditions.

Another novel application of standard primer projects was described by Andersson et al. [9], who used PCRs to create microarray probes. Because they wanted to make microarrays for two bacterial genomes, they needed to design a set of primers so that each primer would amplify products from each genome in order to save on reagent costs. Their approach was to find small regions of sequence agreement between the two genomes, and then use Primer3 to design primers for these regions. They also devised a method for choosing small sets of primers that could amplify regions of both genomes when the primers were used in different combinations.

1.1.3 k-mer Approaches to Primer Design

Some researchers have focused instead on analysis of k-mer distribution in genomic DNA. Their hypothesis was that primers with infrequently occurring k-mers at the 3' end would be specific, and that k-mer distribution in conjunction with a control of GC content in the primer sequence would be sufficient for primer design applications.

Griffais et al. [66] analyzed the distribution of 8-mers in the human genome, and showed experimentally that primers with commonly occurring 8-mers would generate many products in human genomic DNA, whereas primers with rare 8-mers would be more specific. Although their approach specifically amplified a set of bacterial sequences in the presence of human genomic DNA, their approach was not general enough for routine use.

Lopez-Nieto [105], in contrast, used commonly occurring k-mers to amplify a broad range of messenger RNA transcripts. By selecting the most common k-mers in a set of transcripts, they were able to show that they could detect more than 90 percent of long transcripts (with 500 bases or more in the coding region) with only 30 primer pairs. They demonstrated the efficacy of their approach by sequencing clones of PCR products. They also demonstrated that their approach could target conserved gene families (such as G-protein coupled receptors). Their approach had very limited specificity, and was thus not applicable to common primer design problems.

1.1.4 Multiple Target Primer Design

In some experimental situations, primers pairs must be designed that amplify more than one target. One application already described was addressed by Andersson et al., who designed primers to amplify targets in multiple genomes simultaneously; this work was driven primarily by the desire to reduce reagent costs. Another reason to design primers to amplify multiple targets is to obtain sequences for members of a gene family in organisms which have not been studied.

Pearson et al. [139] studied the problem of minimizing the number of primers to amplify a set of targets, and were able to show that the problem is NP-Complete (i.e. computationally intractable). Further, Pearson et al. showed that even approximating the solution (i.e. obtaining a solution close to the best one) was intractable. Nevertheless, Pearson et al. devised a heuristic approach for choosing a small set of primers to amplify a larger set of targets. Hsieh et al. [78] presented a different heuristic solution for an application in which they designed a small set of primers to amplify a set of 32 genes involved in cell development.

A more widely used approach to this problem is the design of so-called degenerate primers. Rather than choosing a small set of primers to amplify a set of targets, an effort is made to design mixtures of primers so that all possible sequence variants are generated. This approach is driven by DNA synthesis technology, in which DNA sequences are synthesized base by base. Rather than adding a single base to a growing sequence, a set of bases can be added to generate a diverse population of DNA molecules. For example, synthesizing the sequence ATGN would result in synthesis of ATGA, ATGC, ATGG, and ATGT.

Rose et al. [160] developed a strategy to design degenerate primers to amplify members of well conserved gene families. Their approach relied on multiple alignments of all members of the family, and they employed a strategy in which the 5' region of the primer represents a consensus sequence and the 3' region is degenerate. This approach allows the 5' region of the primers to anchor primer binding, and the 3' region to exactly match all sequence variants in the multiple alignment. Rose et al. were able to demonstrate successful amplification of previously unknown examples of several genes, such as reverse transcriptase genes in retroviruses and methyl transferases in a wide variety of species, spanning mammals, plants, and invertebrates.

Another example of multi-target primer design is for the application of differential display. In differential display, many messenger RNA products are amplified and visualized on a gel. These bands were used to detect gene expression variation across multiple conditions. Fislage et al. [55] developed a method for choosing primers to amplify messenger RNA transcripts. They chose 10-mer and 11-mer primers, such that the 10-mer primers occurred frequently in coding sequences, and contained a start codon sequence at the 3' end. The 11-mers were also chosen to contain common 10-mers, and contained a stop codon; thus, used together, these primers could be expected to amplify messages with a start and a stop codon, which should contain nearly complete transcripts (except for five and 3' untranslated regions). Their primer pairs generated more than four thousand bands from *E. coli* DNA, and they showed via a limited sequencing study that the products were known messenger RNAs; however, the amplified products did not start and end at start and stop codon.

1.1.5 PCR primer specificity

The specificity of PCR primers was a concern from the very beginning of research into PCR primer design. Very little could be done to address specificity, however, before genome sequences were available: without knowing the DNA sequences in the reaction, it was impossible to predict with certainty whether a primer pair would amplify undesired templates. Once genomic sequences became available, genomes were searched for approximate matches to primer sequences using the BLAST algorithm [8]. This approach could assure that there were no very close matches to the primer sequences that were close in the genomic sequence oriented so that they could generate additional amplicons.

Although BLAST is extremely effective at finding homologous sequences to long queries, it is less sensitive for shorter primer length sequences. Because significant binding can occur even with several mismatches in a primer length sequence, BLAST may miss thermodynamically stable but statistically insignificant matches in the genomic sequence. In addition, the homology based alignments used by BLAST differ in important respects from the alignments used in thermodynamic stability computations [172]. For example, one important difference is that thermodynamic stability depends essentially on dinucleotide content, which is not typically captured in homology scores.

Before genome sequences were available, primer design methods addressed two specificity concerns. First, primers could be screened for similarity to non-priming positions in the sequence to be amplified; this was consistently done from the very beginning [167, 73]. Second, programs such as Primer3 [164] used libraries of known repeat sequences to ensure that primers would not have matches to repeat sequences. Matches to repeat sequences could lead to non-specific reactions, especially in mammalian genomes in which there is significant repeat content [10].

Miura et. al. [124] proposed an approach to assessing stability based on the 3' end of the primer. Using thermodynamic parameters, they defined a threshold based on duplex stability of a segment at the 3' end and primer concentration. Specifically, they identify the shortest sequence at the 3' end that can be expected to provide a significant level of binding, at a specified primer concentration, and then search for occurrences of these short sequences in genomic DNA. They were able to show that this approach yielded specific primers in their application.

Another approach available in the UCSC genome browser [90] is based on similarity to the 3' end. In this application, the user must select the number of bases considered at the 3' end of the primer, defining a short region of exact agreement and a longer region of good agreement; these regions are the same for both primers. This approach has the limitation that it does not allow for indels in the hits to the primer sequences and doesn't explicitly account for the stability of the 3' end of the primer. In contrast, Schuler [178] in a similar application allows the user to set the number of mismatches and indels used to search for matches to the primer sequence.

1.2 Polymerase Extension of Mismatches At the 3' Terminus

Any method that aims to predict PCR primer specificity should incorporate known polymerase behavior when there are mismatches at the 3' end of the primer to the template. Because a high mutation rate would be deleterious to most organisms, polymerase enzymes have evolved mechanisms to ensure that DNA replication proceeds accurately. The thermostable polymerase first used in PCR, the Taq polymerase from *T. aquaticus*, does not have an intrinsic proofreading activity, and so extends primers with mismatches at the 3' terminus to some extent. The degree to which Taq polymerase tolerates mismatches is sequence dependent. Several studies have been done on mismatch extension efficiency for these polymerases.

The polymerase literature contains conflicting data about Taq polymerase fidelity in a PCR context. Kwok et al. [94] studied the effect of terminal 3' mismatches on PCR amplification efficiency. They found that of all the possible mismatches, A:G and C:C mismatches reduced efficiency of PCR amplification by a factor of approximately 100. A:A mismatches reduced efficiency by a factor of 20, and all other mismatches were efficiently amplified. They also found that two mismatches in the final four bases at the 3' end were sufficient to disrupt amplification, unless one of those bases was a T.

In contrast, Simsek et al. [183] studied G:T, G:A, and G:G mismatches at the 3' end of the primer, and found G:T mismatches were efficiently extended, whereas G:A and G:G mismatches did not produce PCR products. These results are only partially consistent with Kwok et al., with discrepancies with respect to the efficiency of synthesis of primers with 3' G:G mismatches.

The degree to which mismatches are tolerated near the 3' end is determined by the polymerase. For Taq polymerases, it seems clear that some amount of mismatch at the 3' end of the primer are tolerated; however, many polymerases have been engineered for PCR applications that are much more stringent. These stringent polymerases extend mismatched 3' termini much less effectively.

A primer design method can take three approaches to incorporating information about polymerase fidelity. It can take a conservative approach, in which the worst case polymerase behavior is assumed, and reject primers if there are any sequences that could possibly generate nonspecific products assuming an enzyme like Taq polymerase (with no proofreading activity). It can take a tailored approach, in which it uses information about the specific polymerase employed in a PCR, and adjusts the similarity criteria accordingly. Finally, it can take an optimistic approach, in which proofreading activity is assumed and only look for nonspecific amplification of sequences with exact matches at the 3' end of the primer. The third approach, which was employed to design primers for the empirical test of the primer design methods described later, works well in practice. However, in chapter 3, I also describe a much more sensitive method to identify DNA binding sites that can find many more potential binding sites at a higher computational price.

1.3 DNA Stability

Chemists devoted sustained attention to the problem of predicting DNA binding stability. This field is of crucial importance for primer design, because PCR relies fundamentally on DNA binding and folding behavior.

Broadly, researchers studying DNA stability focused on two research goals . The first is a systematic effort to determine the stability of DNA structural motifs, such as base pairing, base stacking, isolated internal mismatches, and loops (regions of unbound DNA in one or both strands). The second is a long term effort to employ statistical mechanical methods to use these parameters in a principled way to predict the stability of binding between two arbitrary molecules.

In addition to these two major areas of research into DNA stability, there has also been a systematic effort to extend these approaches in order to predict the behavior of DNA in a range of solvent conditions. Because the concentration of cations and non-polar solvents has a strong effect on DNA stability, significant work has been devoted to understanding how varying salt concentrations affect stability [3]. This work is important in a PCR, as the salt conditions can vary significantly from those used in the study of DNA stability.

The nearest neighbor models, formulated by Gray and Tinoco [65], quickly became the standard paradigm for work in DNA stability. In this approach, the sequence dependent properties of polymers are decomposed into contributions by overlapping subsequences. For example, in a dinucleotide model, the stability of a sequence ATGC would be decomposed into the sum of stabilities of AT, TG, and GC. Because the stability of DNA is mostly due to hydrogen bonding between bases and interactions between neighboring carbon rings in DNA bases, a dinucleotide model explains the stability of DNA well [22, 83].

Although a number of approaches were employed to determine these dinucleotide parameters [11, 202, 47, 189, 203], Santa Lucia [171] showed that many of the approaches used were in broad agreement. Santa Lucia and coworkers then steadily increased the scope of the nearest neighbor model, measuring stability parameters for such features as single internal mismatches [4, 143, 7, 6, 5], and dangling ends [20]. The collected set of parameters, in addition to stability parameters for loops and salt corrections to duplex stability, are collected in a 2004 review article [174]. These parameters are used in the work described in this thesis.

Even with a set of accurate stability parameters, their application to the problem of predicting binding stability in the general case was still difficult. There are many potential arrangements in which two molecules can bind via hydrogen bonding, and accounting for all of these conformations is difficult without the development of efficient techniques for enumerating them. Building on a statistical mechanical model of DNA denaturation [146], both Dimitrov et al. [45] and Garel et al. [60] proposed general methods for computing DNA binding stability independently in 2004. In these approaches, dynamic programming is used to compute the stability of a wide variety of conformations in which one strand is bound to the other via hydrogen bonding, and the overall level of binding stability can be deduced from these calculations. These approaches are limited in that they neglect conformations involving intra-strand hydrogen bonding or pseudo-knots (these are conformations which can not be represented as contiguous segments of helix and loop); however, even if these structures could be efficiently enumerated by dynamic programming algorithms, the required stability parameters for pseudo-knotted structures are not available.

Another important consideration is the stability of the possible secondary structures adopted by a DNA molecule. If a primer is stably folded, then it may be unavailable for binding to the template. McCaskill [115] developed a statistical mechanical approach that considered all possible folded configurations (again, except those involving pseudo-knots), and could thus predict the stability of secondary structures. In conjunction with the DNA stability parameter set [174], this approach can be used to compute the free energy of DNA folding.

As of 2004, principled methods and data were available to accurately compute the stability of the major DNA interactions of concern in a PCR. Thus, the question of whether these approaches are useful in primer design can be readily addressed.

1.4 Outline of Thesis

This thesis begins with a description of DNA binding models in chapter 2. These models underlie this work, and are described in detail.

After describing thermodynamic models of DNA binding stability, a method for searching for primer binding sites is described in chapter 3. In contrast to homology based searches, the method is based on thermodynamic principles and shows high sensitivity for primer binding site search.

In chapter 4, a new primer design algorithm is presented. In this approach, the free energies for a number of possible DNA binding interactions in a PCR are computed using thermodynamic models and data, and a simple criterion for primer pair acceptability is described. Essentially, primers are accepted if the concentration of primers binding to template at equilibrium, in the presence of competing binding and folding interactions, is above a user specified threshold.

Finally, in chapter 5, a method is described for designing primers to amplify multiple targets when the sequences are entirely unrelated. This approach is based on a general optimization approach coupled with a method for thermodynamic outlier rejection.

I finish with a chapter describing the contributions and conclusions of this thesis work, as well as a description of future work and open problems.

Chapter 2

DNA BINDING MODELS

2.1 Introduction

Many of the most useful methods in experimental biology rely on DNA binding interactions. Examples of these extremely useful methods are microarray analysis, the polymerase chain reaction, and DNA sequencing. These methods require that binding reactions be stable enough for the intended experimental purpose, and the binding reactions must also be specific in order to prevent undesired hybridization from obscuring the desired measurements.

This practical importance has motivated sustained research into the chemistry of DNA binding interactions [83, 62, 13, 202, 92, 203, 19, 44, 89]. Much of the work has been focused on obtaining detailed stability data that allows the free energy of binding between complementary DNA molecules to be predicted with good accuracy.

Another avenue of research focused on analysis of the partition function for DNA binding [146, 60, 45]. These partition function approaches can compute the stability, or binding affinity, between two arbitrary DNA molecules. The partition function incorporates the stabilities of all conformations in which one DNA molecule could bind to another with at least one base pair. The number of such conformations is comparable to the number of alignments between two sequences, and like alignments, there are too many conformations to explicitly enumerate.

In order to calculate the binding affinity between DNA molecules, the Generalized Poland-Scheraga (GPS) [146, 60, 45] models employ dynamic programming methods to account for each of a restricted class of conformations. Because of the systematic neglect of particular kinds of conformations (such as intramolecular hair-pinning and pseudo-knots), the GPS models are limited in accuracy, especially for longer molecules.

In addition to DNA binding, DNA folding (where bases in a single DNA strand bind to one another via hydrogen bonding) is also of practical importance; in PCR primer design, a primer with a stable folded form would fail as a primer because the folded form could eliminate its ability to bind to its priming target. In an approach similar to the dynamic programming methods used to predict binding affinity, dynamic programming is also used to enumerate the ways in which a single molecule can fold [115]. These methods largely use the same parameters as the DNA binding affinity calculations, and have similar limitations.

2.2 Generalized Poland-Scheraga Models

The DNA binding partition function integrates the ensemble of ways that two molecules can bind into a final measure of binding affinity. The binding affinity, K, is the ratio, at thermodynamic equilibrium, of the concentration of the dimer form of DNA (in which the two strands are bound) to the product of concentrations of the unbound forms. For two sequences X and Y, let $[XY]$ represent the concentration of the dimer form, and $[X]$ and $[Y]$ represent the concentrations of the unbound single stranded forms in units of moles per liter. A quantity called the Gibbs free energy of the binding reaction (ΔG) is related to the binding affinity as

$$\frac{[XY]}{[X][Y]} = K = e^{-\frac{\Delta G}{RT}}, \tag{2.1}$$

where R is the molar gas constant and T is the temperature in degrees Kelvin. When the ΔG of a reaction is negative, the reaction is thermodynamically favorable, and the concentration of products will be high at thermodynamic equilibrium. When the ΔG is positive, then the reaction is thermodynamically unfavorable, and the concentration of reaction products will be low at thermodynamic equilibrium.

A

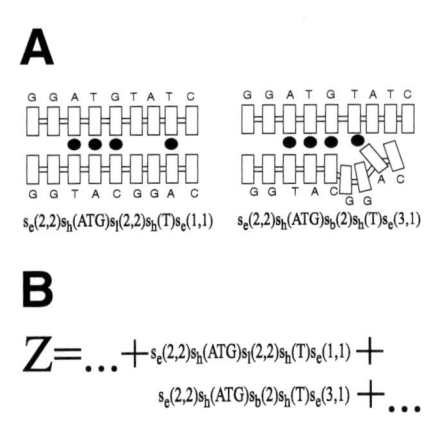

G G A T G T A T C
G G T A C G G A C

$s_e(2,2)s_h(ATG)s_l(2,2)s_h(T)s_e(1,1)$

G G A T G T A T C
G G T A C
G G

$s_e(2,2)s_h(ATG)s_b(2)s_h(T)s_e(3,1)$

B

$$Z = \ldots + s_e(2,2)s_h(ATG)s_l(2,2)s_h(T)s_e(1,1) + s_e(2,2)s_h(ATG)s_b(2)s_h(T)s_e(3,1) + \ldots$$

Figure 2.1: Nucleic acid binding affinity. The Generalized Poland-Scheraga models of DNA binding express the binding affinity between two molecules as the sum of stabilities of all conformations in which there is at least one base pair (represented with a dot) between the two molecules. (A) The stability of a conformation is decomposed into a product of terms accounting for the stability of unbound ends, helical regions, internal loops (left), and bulge loops (right). (B) The partition function, Z, is the sum of stabilities of conformations with at least one base pair.

The GPS models of nucleic acid binding express the total binding affinity between two nucleic acid molecules as the sum of binding affinities for each conformation where one molecule has at least one base pair with the other. Figure 2.1A shows two such conformations. The stability of a conformation is decomposed into stabilities of unbound ends, unpaired bases (called loops), and bound regions. The total stability of a conformation is the product of the stabilities of its constituent parts. The stability of the helical regions depends strongly on the dinucleotide composition. In contrast, for DNA, the dependence of loop stabilities on sequence is largely unknown, and loop stability is frequently taken to be independent of sequence.

The stabilities of all conformations with at least one base pair are summed together to get the total stability. The sum of all stabilities, Z, is referred to as the partition function as depicted in Figure 2.1B. In this work, the partition function Z is taken to be equal to K, i.e.

$$\Delta G = -RTln(Z) \qquad (2.2)$$

This relationship results from using the free energy of the unfolded strands as the reference state. Although base stacking occurs in single strands and can result in temperature dependent values for the change in enthalpy and entropy of DNA binding [45], this temperature dependence is neglected in this work.

2.2.1 Conformations of DNA binding

For two bound strands of DNA, a conformation describes the arrangement of hydrogen bonds between the strands. The Generalized Poland-Scheraga models of DNA binding enumerate a subset of the possible conformations of DNA binding. In particular, they neglect conformations in which there is intramolecular base pairing (such as conformations with hairpin loops), as well as conformations incorporating pseudoknots. This section describes the set of conformations considered by the GPS models of DNA binding.

Let a conformation of binding between two DNA sequences X and Y be an ordered list of ordered pairs (a_i, b_i). In particular, if there is a base pair between the ith base of X and the jth base of Y, then the pair (i, j) will be in the set of pairs associated with the conformation. This list is sorted in order of the first coordinate, so that $a_i < a_{i+1}$.

Figure 2.2 shows two conformations and the set of ordered pairs specifying each conformation. The conformation in figure 2.2A is of the type accounted for by GPS models; in contrast, the conformation in figure 2.2B is of the type that is not considered by GPS models.

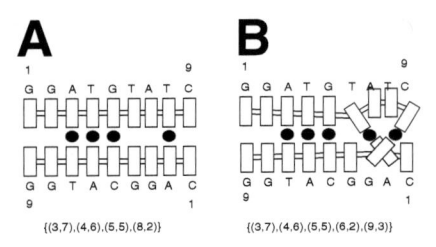

Figure 2.2: Two conformation of a DNA dimer. (A) An allowed configuration, in which the sequences are bound. (B) A forbidden conformation with a pseudo-knot.

The Generalized Poland-Scheraga models of DNA binding consider all conformations in which there are no hairpins and no pseudo-knots. These restrictions can be expressed as constraints on the set of conformations considered. In particular, a conformation C is admissible if:

$$\forall a_i, b_i, a_j, b_j \in C, a_i > a_j \rightarrow b_i < b_j; \tag{2.3}$$

$$\forall a_i, b_i \in C, X_{a_i} = C(Y_{b_i}), \tag{2.4}$$

where $C(Y_i)$ is the Watson-Crick complement of the ith base of Y.

Each conformation can be decomposed into unpaired ends, helical regions, and internal loops. The stability of each of these regions can be computed using reference data as described below.

2.2.2 Stability factors and free energy

I use *stability factors* to quantify the stability of particular features of two bound DNA strands, such as base pairs and loops. Stability factors are similar to free energies, except they are exponentiated so that multiplying two stability factors is equivalent to adding free energies.

The stability factor s associated with a free energy ΔG is

$$s = e^{\frac{-\Delta G}{RT}}. \tag{2.5}$$

When two stability factors are multiplied, the resulting stability factor is associated with the sum of the free energies:

$$s_1 s_2 = e^{-\frac{\Delta G_1}{RT}} e^{-\frac{\Delta G_2}{RT}} \tag{2.6}$$
$$= e^{\frac{-(\Delta G_1 + \Delta G_2)}{RT}}. \tag{2.7}$$

A negative ΔG implies a thermodynamically favorable reaction. If the ΔG of a reaction is negative then the associated stability factor will be greater than one. Conversely, an unfavorable reaction will have a positive ΔG and the associated stability factor will be less than one; thus, stability factors associated with favored reactions will be large, and stability factors associates with unfavored reactions will be small.

The ΔG of a reaction is determined by the change in enthalpy ΔH and the change in entropy ΔS of the reaction. The change in enthalpy is a measure of the heat energy gained or lost in a chemical reaction, and the change in entropy is a measure of the overall disorder. The Gibbs energy is related to these parameters as

$$\Delta G = \Delta H - T\Delta S. \tag{2.8}$$

Reference data specifies the entropy and enthalpy for each binding motif, and the free energies are computed for specific temperatures using these values.

2.2.3 Helical Stabilities

The stability of a helix depends on hydrogen bonding and so-called base stacking interactions. The most accurate models of helix stability are the nearest neighbor models [65, 83], which decompose the stability of a DNA helix into the product of stabilities of all consecutive pairs of bases. These models also include a helix initiation parameter, that must be included to account for the thermodynamic energy of initiating a DNA helix, as well as a penalty for closing a helical segment with an AT base pair. Using the nearest neighbor model, the stability of the helix with sequence ATGC would be computed as

$$s_h(ATGC) = s_i s_e(A/T) s_h(A,T) s_h(T,G) s_h(G,C), \qquad (2.9)$$

where s_h is a stability factor associated with a helical region of two bound DNA molecules and s_i is the stability factor that accounts for the initiation penalty. The factor $s_e(A/T)$ is the penalty for closing with a terminal AT base pair in order to account for end effects.

2.2.4 Loop Stabilities

Loops are thermodynamically unfavorable and have small stability factors. There are two kinds of loops: bulge loops and internal loops [83]. See the conformation in Figure 2.1A on the right for an example of a bulge loop. A bulge loop is analogous to an insertion in traditional alignment algorithms, and contains unpaired bases on only one strand. An internal loop, by contrast, contains unpaired bases on both strands. An example of an internal loop is shown in the conformation on the left in Figure 2.1A.

The free energy of a bulge loop of length l [83] is extrapolated from a known and measured stability,

$$\Delta G(l) = \Delta G(n) + rRT\ln\left(\frac{l}{n}\right), \qquad (2.10)$$

where $\Delta G(n)$ is the stability of a bulge loop of length n (which is measured experimentally), R is the molar gas constant (1.987 calories per mole), T is the temperature in degrees Kelvin. At 50C, the temperature at which the reference data is most accurate, RT is approximately 650. The constant r is a parameter that is usually taken to be 1.5 from theoretical considerations or 2.44 from experimental measurements; the experimentally derived value is used here. There is some choice in the value of n, from which free energies are extrapolated.

The stability of a bulge loop of length l is

$$s_b(l) = \frac{\omega_0}{l^r} = e^{\frac{-\Delta G(l)}{RT}}, \qquad (2.11)$$

Where ω_0 is a constant accounting for the factor of RT and the reference free energy. I use experimentally measured stability values for short bulge loops and the extrapolations to longer bulge loops as presented in [83].

The free energy of an internal loop [83] with l_1 unpaired bases in one strand and l_2 bases in the other strand (where both l_1 or l_2 are greater than 1) is

$$\Delta G(l_1, l_2) = \Delta G(n) + rRT\ln\left(\frac{l_1 + l_2}{n}\right) + c\mid l_1 - l_2 \mid \qquad (2.12)$$

where n is the value for a particular reference lenght, and the constant c is a penalty for asymmetry in the internal loop, and is about 300 calories per mole.

As with bulge loops, the free energy for internal loops is extrapolated from the measured free energy of a particular length. The stability of an internal loop with l_1

unpaired bases in one strand and l_2 bases in the other strand is

$$s_l(l_1, l_2) = \frac{\omega_1}{(l_1+l_2)^r} e^{-\frac{c|l_1-l_2|}{RT}}, \qquad (2.13)$$

where r is as before and ω_1 is derived from an experimentally measured loop stability. As with bulge loops, I use extrapolations in [83].

For internal loops of length two, however, comprehensive data are available to account for sequence dependent loop stabilities [83]. These stability factors are used in preference to the extrapolated values for isolated internal mismatches.

The reference data for bulge and loop stability are a mixture of experimentally measured values and extrapolation for unmeasured values. However, the errors due to extrapolating from a single value are thought to be fairly small [83], and hence extrapolations are reasonably accurate.

2.2.5 Unpaired End Stabilities

The stability of unpaired ends is simply the product of factors accounting for the stability of unbound bases at the end each strand [60]. If there are l_1 unpaired bases at the 5' end of the top strand, l_2 unpaired bases at the 3' end of the top strand, l_3 unpaired bases at the 5' end of the bottom strand, and l_4 unpaired bases at the 3' end of the bottom strand, then the stability factor due to unpaired bases is

$$s_e(l_1, l_2) = \omega_2^{l_1} \omega_2^{l_2} \omega_2^{l_3} \omega_2^{l_4}. \qquad (2.14)$$

where ω_2 is a measured experimental value.

Experimental work has shown that unpaired strands dangling from helical segments do not contribute to dimer stability, although there is a stability contribution from the mismatch adjacent to the final helical segments[83]. However, this detailed stability data remains unpublished, and we do not use it. We therefore use a value of

1 for ω_2.

2.2.6 The stability of a conformation

Given a conformation, and two DNA sequences X and Y, the stability factor for the conformation is

$$S(X,Y,c) = S_h(X,Y,C)S_l(X,Y,C)S_e(X,Y,c), \qquad (2.15)$$

where $S_h(X,Y,c)$ is the product of stability factors for all helical segments of the conformation, $S_l(X,Y,c)$ is the product of stability factors for all loops, and $S_e(X,Y,c)$ is the product of stability factors for unpaired ends.

2.2.7 Accuracy of the reference data

The thermodynamic reference data is most accurate at 50 C [83]. One limitation of the existing data is an assumption that the entropy and enthalpy do not change with temperature; however, a number of reports [195, 31, 194] concluded that the existing parameter set does not capture an important temperature dependence. The extent to which this temperature dependence may bias calculations that depend on the existing reference data has not been systematically evaluated.

Some investigators examined the mechanisms for this temperature dependence [121, 122, 74] and concluded that the temperature dependence of is due primarily to the amount of base stacking in unbound strands. At low temperature, base stacking in unbound strands is significant, and the enthalpy change upon binding is less than at high temperature. At high temperature, there is much less base stacking, and the change in enthalpy is greater. However, base stacking in single strands is sequence dependent [74, 122], and reference parameters for base stacking in single strands have not been published. The approaches in this work rely on the ability to predict stability at different temperatures, and they will all be significantly more accurate when improved reference data becomes available.

The concentration of cations in solution affects DNA binding stability, as well as a number of additives used in PCRs. In addition, PCR conditions have salt concentrations that differ significantly from those salt conditions used to measure the reference data (50mM NaCl in typical PCRs as compared with 1M NaCl in typical experimental studies of DNA stability). Thus, extrapolations to relevant salt conditions are necessary to adjust the final free energy. Various empirical extrapolations have been presented [83, 3], which differ to some extent.

2.2.8 The Partition Function

The partition function for sequences X and Y is the sum of stabilities associated with all members of the set C of admissible configurations

$$Z(X,Y) = \sum_{c \in C} S(X,Y,c). \tag{2.16}$$

The dynamic programming algorithm used in this work consists of two steps. The first step fills a dynamic programming matrix with the stability factors for partially bound sequences, and the second step integrates this information to compute an overall stability. Detailed derivations and presentations of these dynamic programming algorithms are given in [60, 45]. For prediction of DNA binding, I use the dynamic programming approach presented in [45], and for prediction of DNA folding, I use the dynamic programming approach presented in [46] to compute folded stability without considering pseudo-knots.

Chapter 3

EFFICIENT IDENTIFICATION OF DNA BINDING PARTNERS IN A SEQUENCE DATABASE

3.1 Introduction

Many[1] fundamental methods in molecular biology rely on binding between complementary DNA molecules. For instance, the polymerase chain reaction (PCR) [170] relies on the specific binding of short DNA primer sequences to the DNA of interest. PCR is used in a multitude of contexts [82], from disease diagnosis [86] to gene expression measurement [216]. DNA microarrays [175] also rely on the specific hybridization of array probes to DNA sequences in a mixture in order to measure gene expression or determine sample genotypes [188].

Assays that rely on hybridization are compromised when primers or probes bind non-specifically to DNA molecules that are not their targets [33]. In the presence of non-specific hybridization, measurement accuracy in quantitative assays can be severely compromised, especially when the hybridization target is present in low abundance. Even in the context of non-quantitative PCRs, non-specific binding can lead to the formation of undesired products that compete with the reaction of interest and reduce reaction yields. Therefore, assessing hybridization specificity is an important part of the design of these reactions.

The most straightforward approach to assessing hybridization specificity would be to query every potential binding site in the background DNA for binding affinity. In

---

[1]Portions of this chapter were previously published in the Proceedings of the Intelligent Systems for Molecular Biology conference in 2006 [110].

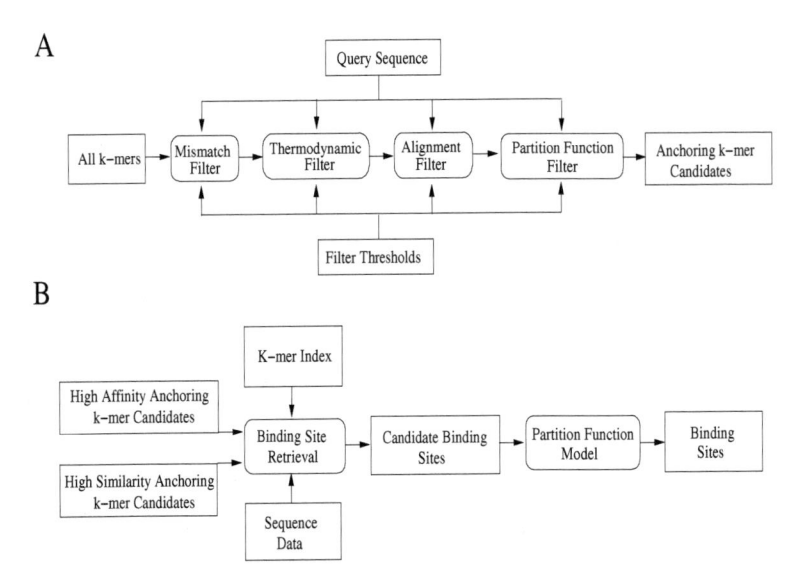

Figure 3.1: Overview of filtering algorithm. **(A)** k**-mer filtering.** All k-mers for a specified value of k are input to the mismatch filter, along with a set of pre-chosen similarity thresholds. The four filters eliminate k-mers in turn, producing as output a list of candidate k-mers that could anchor a binding site. We subject all k-mers to two sets of thresholds, producing two sets of candidates binding site anchors. One set yields k-mers that have high thermodynamic affinity to the query, and the other set yields k-mers that have high sequence similarity to the query. **(B) Candidate retrieval and evaluation.** The k-mers that passed the filtering steps in (A) are located in the genome sequence using a precomputed index. We examine only those sites where a candidate k-mer from one group occurs with close proximity to a candidate k-mer from the other group. These candidate binding sites are then tested for binding affinity using the partition function model, and all sequences that bind to the query with greater than a target affinity are reported.

most experiments, the background DNA that comprises the reaction mixture consists of the genome of the organism being studied. Hence, for the human genome, this approach requires evaluating approximately six billion possible binding sites, corresponding to the two strands of each chromosome.

In practice, applying state-of-the-art DNA binding models on a genomic scale is not computationally feasible. These models use dynamic programming algorithms with a computational complexity of $O(nm)$ for two sequences of length m and n, respectively [60, 45], and the complexity of querying an entire genome is $O(gmn)$, where g is the number of bases in the genome, m is the sequence length, and n is the size of the genomic subsequence queried at each position. In our experiments, scanning the complete human genome for binding sites to a 25-mer probe requires approximately 180 days of CPU time. For most primer or probe design applications, this is clearly too long to wait.

Current practical methods for predicting non-specific binding of a given DNA sequence rely on heuristic approximations. Perhaps the most commonly used method for identifying binding sites between a query DNA sequence and a target genome predicts binding sites based upon a pre-specified maximum number of mismatches between the probe's reverse complement and the target [90, 106, 209, 220]. As we demonstrate below, this approach is inaccurate because sequences can stably bind in the presence of bulge loops, which correspond to insertions and deletions in an alignment. An alternative method for identifying non-specific binding sites relies on the BLAST algorithm or other alignment based criteria [8, 69, 221, 9]. This approach, too, is inaccurate, primarily because BLAST is designed to detect statistically significant sequence homology, rather than sequence binding partners.

We propose a filter- and index-based method, shown in Figure 3.1, for rapidly identifying binding partners of a given query sequence. In the initial stage (A), we consider all possible k-mers of a given length and identify k-mers that could anchor a binding site to the query sequence. This stage includes four filters that are designed to recognize various aspects of DNA binding stability. Two of the filters were developed for this application. The filters are applied in order of increasing computational complexity, so that most k-mers are excluded by the simplest filters. Using our approach and considering 10-mer anchors, typically more than 99% of target 10-mers are excluded from further consideration. In stage (B), we use a suffix array index to rapidly extract the sequence context of all occurrences of the k-mers obtained in the first step. These candidate binding sites are then evaluated using a model of DNA binding. Because so many k-mers are excluded at the outset, we can afford to apply an accurate model of DNA binding in the second stage of the algorithm.

Using our method, we achieve rapid and comprehensive identification of likely binding sequences. The first stage of the algorithm reduces the sequence search space by three orders of magnitude. The second stage is quick because many of the occurrences of the k-mers that pass the filtering stage can be eliminated by further filtering. Furthermore, our filter thresholds are set to achieve this speedup while retaining 100% accuracy, compared with considering every possible binding site in the target genome. Our approach reduces the amount of time to scan a sample 30 MB sequence from two days to under a minute for typical queries.

3.2 Algorithms

We hypothesize that binding sites in genomic DNA can be comprehensively retrieved by first identifying short regions of agreement between the query sequence and the genomic DNA, and then examining the sequences containing these short regions of agreement with accurate models of DNA binding. We base this hypothesis on the observation that the thermodynamic instability of unbound bases in a DNA duplex limits the amount of disagreement between a query sequence and any of its binding sites.

In particular, our method relies on a set of filters to identify k-mers that have good agreement with the query sequence, and could therefore anchor a binding site.

In this section, we describe state-of-the-art models of DNA binding and then explain how our filters relate to those methods.

3.2.1 Partition function models of DNA binding

The overall goal of a model of DNA binding is to predict the *binding affinity* of a given pair of DNA sequences. The binding of two single stranded DNA molecules to form a dimer is a reversible reaction, and the binding affinity reflects the balance of association and dissociation reactions in a large population of molecules at thermodynamic equilibrium. When the binding affinity is large, then the dimer form is favored, and when the binding affinity is small, then the single stranded forms are favored.

Recently, efficient dynamic programming methods have been developed to compute the affinity of two DNA molecules [60, 45]. In this approach, a dynamic programming algorithm computes the sum of the exponentials of the energies of almost every alignment in which one molecule has at least one base pair with the other molecule. This sum is then proportional to the binding affinity. In this work, we use the HYBRID software [45], which implements one such dynamic programming algorithm. However, our method does not rely on the specifics of the HYBRID software: our filters are designed to account for known, generic features of DNA affinity, and other models of DNA binding could be used in the final step to evaluate the filtered list of candidates. Indeed, although HYBRID and similar methods represent the state of the art in determining the affinity of two DNA sequences, they are known to systematically neglect some alignments that are important in some contexts.

3.2.2 An efficient algorithm for finding binding sites

Our goal is to identify all of the sequences in a database that bind to a query sequence according to a given partition function model of DNA binding. We do this in two stages, as described in Figure 3.1. First, we identify two groups of k-mers. One group of k-mers consists of k-mers with high sequence similarity to the query, and the other group of k-mers consists of k-mers whose reverse complements have high thermodynamic affinity to the query sequence. Each group is defined as the set of k-mers that pass through a series of four filters described below; both groups are passed through the same filters but each group is identified by the use of different filter thresholds for each filter. In the second stage, each location in the sequence where there is a k-mer from the high affinity group within a pre-specified distance of a k-mer from the high similarity group is retrieved, along with flanking sequence. These candidate binding sites are then evaluated using the partition function model. The output of the algorithm is a list of binding partners for the query sequence.

In the first stage of our approach, we consider all k-mers for a prespecified k, and we use a series of four filters to eliminate k-mers that have little affinity or similarity to the query. Each filter is designed to reject those k-mers that have little affinity to the query, and thus restrict the number of candidate binding sequences that must be considered. Furthermore, the filters are designed to be increasingly stringent, and are applied in order of increasing computational complexity; the first filter is very fast but will pass some k-mers with low affinity to the query, whereas the last filter is more expensive to compute but will reject all those k-mers with little thermodynamic affinity to the query. Each filter must be applied in conjunction with a threshold. The threshold for each filter is determined empirically by examining characteristics of binding sites predicted by the partition function model of DNA binding. These thresholds are chosen conservatively, so that each filter will pass some k-mers with low affinity to the query rather than discard k-mers that could anchor a binding site.

Each filter uses a function designed to compare two k-mers. In order to compare a candidate k-mer to a query sequence, we first decompose the query sequence into k-mers of the same length as the candidate k-mer, and then compare the candidate k-mer to each k-mer derived from the query (see Figure 3.2A). If any of the query derived k-mers meet the specified similarity to the candidate k-mer (Figure 3.2B), then the candidate k-mer is retained for further analysis. If none of the query derived

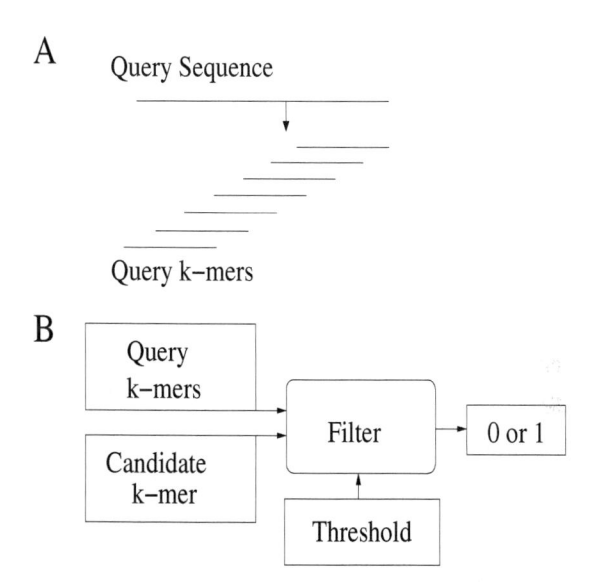

Figure 3.2: Filtering k-mers. **(A) Decomposition of the Query sequence into k-mers.** The query sequence is decomposed into overlapping k-mers of a specified length. **(B) Computation of the similarity of a k-mer to the query.** Each filter identifies k-mers that could anchor a binding site, taking as input the k-mers derived from the query sequence, a candidate k-mer, and a pre-specified filter threshold. Each filter then reports whether the candidate k-mer had the specified level of similarity to at least one of the k-mers in the query sequence or not.

k-mers meets the specified similarity, then the candidate k-mer is eliminated from further consideration.

The simplest filter—the mismatch filter—eliminates k-mers that differ from every k-mer in the query sequence by more than a specified number of bases. This filter is designed to reject k-mers that have little affinity to any part of the query sequence. The filter function computes the fraction of mismatches between a candidate k-mer K and the query sequence Q:

$$F_1(K,Q) = \max_{j \in s(Q,k)} \sum_{i=1}^{k} \frac{\delta(K_i = j_i)}{k},$$

where $s(Q, k)$ returns the set of all k-mers in Q, and δ is the Kronecker delta function.

The second filter rejects k-mers that contain destabilizing internal mismatches relative to the query. These destabilizing mismatches are identified using thermodynamic data on DNA binding stability [174]. This filter's function is similar to the mismatch filter, except that it takes into consideration the specific stabilities of dinucleotide stacks (pairs of adjacent, paired bases) and single internal mismatches. We implement this filter by encoding each k-mer K as a complex valued vector $\Phi(K)$, and we developed this filter so that the inner product of the conjugate of the encoding of one k-mer and another k-mer approximates the sum of the free energy of binding between the first k-mer and the reverse complement of the second k-mer, and vice versa. Details of this encoding are given in Appendix A. The final value of this filter is a normalized dot product:

$$F_2(K,Q) = \max_{j \in s(Q,k)} \frac{\langle \Phi(K), \Phi(j) \rangle}{\sqrt{\langle \Phi(K), \Phi(K) \rangle \langle \Phi(j), \Phi(j) \rangle}}.$$

The third filter rejects k-mers that do not have good sequence agreement with the query, considering the possibility of asymmetric internal loops. For each candidate k-mer, this filter's function considers many alignments with respect to the query sequence, weighting each by the number of matches and the length and topology of loops. Asymmetric internal loops serve to separate regions of sequence agreement, and thus this filter will recognize sequence similarity even when regions of sequence agreement are separated by insertions or deletions in one sequence with respect to the other. We developed this filter function to be a coarse approximation of the partition function for one sequence binding to the reverse complement of the other, and we therefore consider only base pairing (and neglect the detailed thermodynamic reference data on dinucleotide stability) and internal loops of length three or less. In addition, we use loop stability values optimized for this application. The final value of the alignment filter is

$$F_3(K,Q) = \max_{j \in s(Q,k)} \frac{f(K,j)}{\sqrt{f(K,K) \cdot f(j,j)}}.$$

The alignment function $f(\cdot,\cdot)$ is described in appendix A.

The fourth filter applies the partition function model directly. In this step, we compute the binding affinity between the reverse complement of the k-mer and the query sequence. In order to normalize out the binding properties of the query sequence, we divide this binding energy by the binding energy of the k-mer in the reverse complement of the query sequence with the highest affinity to the query sequence. The final value is

$$F_4(K,Q) = \frac{g(\hat{K},Q)}{\max_{j \in s(Q,k)} g(\hat{j},Q)},$$

where a carat denotes reverse complement, and $g(\cdot,\cdot)$ is the partition function model of DNA binding. In practice, this filter is the most stringent and the most computationally complex.

We apply the four filters twice, with two sets of filter thresholds, to get two sets of candidate anchoring k-mers. We do this in order to reduce the total number of candidates that must be considered, by requiring proximity of a k-mer from one group to a k-mer from the other group. We use filter thresholds so that the high similarity group of k-mers will be similar with respect to filters F1 and F2, and the high affinity k-mers will be similar with respect to filters F3 and F4. We then locate all occurrences of both candidate sets in the sequence database, and further consider only those locations in the sequence database where there is a k-mer from the high affinity group close to a k-mer from the high similarity group (see Figure 3.3).

After the four filtering steps, we must efficiently locate all occurrences of the high affinity and high similarity k-mers within the given sequence database. This is accomplished by using a modified suffix array [67, 108] to index the database. In a suffix array, pointers to suffixes of a sequence are sorted lexicographically; in our modified suffix array, the pointers are sorted based on comparison of only the first k positions of the suffix, where k is the length of the filtered k-mers. We also build a hash table on the suffix array itself, so that the positions in the suffix array corresponding to a query k-mer can be quickly located (with a computational complexity of $O(k)$ per k-mer lookup). We use this sequence index, consisting of the modified suffix array and the hash table into the suffix array, to rapidly identify all locations where a candidate k-mer from one group occurs close to a candidate k-mer from the other group. These occurrences, along with their flanking sequences, comprise the list of candidate binding sites.

In the final step, each remaining candidate binding site is evaluated by the partition function model for affinity to the query sequence. As we show in Section 3.4, by using a set of fast, accurate filters, the filtering and indexing stages of the algorithm reduce the sequence search space by three to five orders of magnitude. Therefore, in the final step, we can afford to incorporate a relatively sophisticated, computationally expensive model of DNA stability. Thus, by coupling a pre-filtering step with accurate refinement of the candidate list, we achieve both efficiency and accuracy.

```
Sequence •——————  ··········  ··········  —————→  High Affinity k-mers
         ——————                 ——                High Similarity k-mers
                              |——|
                               W
```

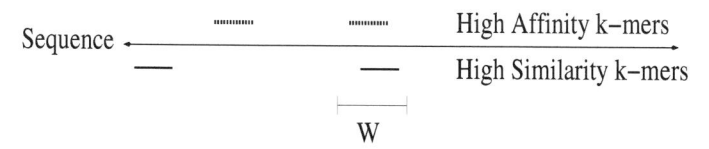

Figure 3.3: Search for proximal hits. Our binding site search algorithm finds anchoring k-mers in the search sequence. We use two sets of filter thresholds, and obtain two sets of candidate anchoring k-mers; one set has high similarity to the query, and the other set has high affinity to the query (occurrences of k-mers from the high affinity set are drawn with dashes above the search sequence, and occurrences of k-mers with high similarity are drawn with solid lines below the search sequence). We locate all occurrences of both groups of candidate anchoring k-mers, and further examine only those sites where there is a candidate anchoring k-mer from the high similarity group occurring within a pre-specified distance w from a candidate anchoring k-mer from the high affinity group.

3.2.3 Choice of filter thresholds

Clearly, the success of our filtering strategy depends to a large extent on the thresholds that we use for each filter. If our thresholds are too stringent, then we risk eliminating true binding partners from our list. Conversely, if our thresholds are not stringent enough, then the efficiency of the search will decrease.

We compute these thresholds empirically by using the partition function model. First, with respect to a given set of experimental conditions and a target level of binding affinity, we scan a sequence database for binding sites to a set of query sequences using the partition function model, storing a list of all binding sites with stability better than a given threshold. We then choose filter parameters conservatively, so that if we re-searched the sequence using our filtering approach, we would obtain all of the binding sites obtained in the slow linear scan.

Our thresholds are set by analyzing the binding sites identified using a linear scan, using the procedure illustrated in Figure 3.4. We decompose each binding site into its constituent k-mers, as in Figure 3.4(A). We then rank these k-mers according

A Query Sequence

Binding Site Sequence

Binding Site k-mers

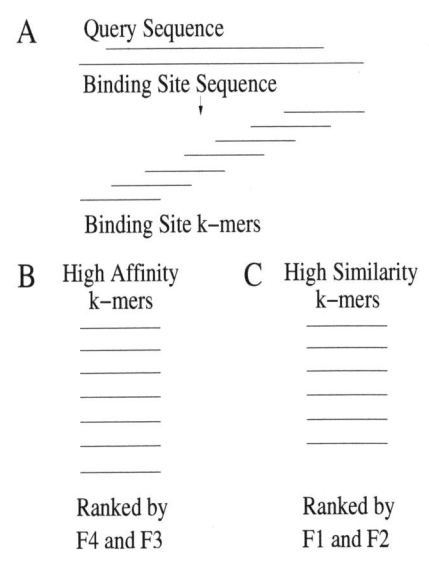

B High Affinity   C High Similarity
   k-mers   k-mers

Ranked by   Ranked by
F4 and F3   F1 and F2

Figure 3.4: Filter Analysis of a binding site. **(A) Decomposition of binding site.** Each binding site is decomposed into its constituent k-mers. **(B) Ranking of binding site k-mers according to thermodynamic affinity.** The binding site k-mers are ranked according to similarity to the query by F4; F3 is used to break ties. The similarity scores of the top ranked k-mer are added to the set of similarity scores used to determine filter thresholds for the high affinity group of candidate k-mer binding site anchors. **(C) Ranking of binding site k-mers according to sequence similarity.** All k-mers, except the top ranked k-mer in (B) are re-ranked according to sequence similarity to the query by F1; F2 is used to break ties. The similarity scores of the top ranked k-mer are added to the set of similarity scores used to determine filter thresholds for the high similarity group of candidate k-mer binding site anchors.

to similarity to the query sequence using the filter functions. For the high affinity group of filter parameters, we rank first by filter function F4 and break ties using filter function F3 as shown in Figure 3.4(B); filters F4 and F3 are designed to identify high affinity k-mers, rather than high similarity k-mers. For the high similarity group of filter parameters, we remove the top ranked k-mer and re-rank the remaining k-mers using the filter function F1; we break ties with filter function F2 as shown in Figure 3.4(C); we use F1 and F2 because these filters identify similarity more than affinity. We then compute the similarity of both top ranked k-mer to the query according to all filter functions.

After analyzing each binding site recovered from the linear scans, we integrate information from all binding sites as follows. Each binding site contributes two sets of similarity scores, one set for the top k-mer ranked according to thermodynamic affinity and one set for the top k-mer ranked according to sequence similarity. We accumulate all sets of similarity scores into two sets. One set contains the similarity scores for all top ranked k-mers according to thermodynamic affinity, and the other set contains the similarity scores for all top ranked k-mers according to sequence similarity. To obtain the final filter parameters, we find the minimum similarity score in a set over all binding sites for each filter function. This is thus a conservative method for obtaining filter parameters, and ensures that if the sequence were re-searched with our filtering approach, we would recover all of the binding sites identified with the linear scan.

Intuitively, the two sets of filter thresholds capture different characteristics of DNA binding: sequence agreement and k-mer binding affinity. These two notions of similarity are not the same: consider a query sequence that consists of several A bases followed by several G bases. A k-mer consisting entirely of A bases would have perfect sequence agreement to the left part the the sequence, whereas a k-mer consisting of all G bases with two consecutive internal A bases would have poor sequence agreement, but the reverse complement of that k-mer would have much higher binding affinity to the query sequence than the reverse complement of the sequence consisting of all

Table 3.1: ΔG PCR: Free energy of binding, in kilocalories per mole, of the sequence to its reverse complement at 55 C in 50 mM NaCl and 2 mM MgCl$_2$; ΔG MA: Free energy of binding, in kilocalories per mole, of the sequence to its reverse complement at 40 C in 1 M NaCl. Energies are computed using the HYBRID software.

	Query sequence	Length	GC	ΔG PCR	ΔG MA
1	GAGCTGCGGCAGAGGCTGGCGCCC	24	0.79	-24.5	-36.8
2	GCCTGCACTGGCTTCAGGAAGCTGGAGCC	29	0.65	-25.3	-40.1
3	GGCCAGTTCCTGCAGCCCGAGGC	23	0.74	-21.6	-33.2
4	AGTGGCATGCCTCTCTCTACCCAGC	25	0.60	-19.7	-32.2
5	CCACCAAAAAGTAATTAAAGGGTTTGCCTCAT	32	0.38	-19.5	-35.6
6	CACGCAAATCATCCCCAGCCACATC	25	0.56	-19.1	-31.8
7	CAGGTGTCCCTGCTTCGGCTTCCAG	25	0.64	-20.6	-33.3
8	CGCGAAGTGACCTTCAGAGAGTACGCCAT	29	0.55	-22.3	-37.2
9	CTGGACTGCCAAGTCCAGGGCAGGCC	26	0.69	-23.0	-36.1
10	GTCACCCACCTGCTGGCCCCGG	22	0.77	-20.9	-32.0
11	GGGGCTCAATAAGTCTGCTTCCACCTT	27	0.52	-19.5	-33.0
12	GGGTGAGGCCCATTCATAAGACTGGC	26	0.58	-19.6	-32.7
13	CCAGTCATGTTGCCCCGTTTGTCAGAG	27	0.56	-20.4	-34.1
14	GGGAGGGCTGAAGAGGGCACTCC	23	0.70	-19.4	-30.9
15	GGATGCATATGGACTCTTAGGTGTTCTGCG	30	0.50	-20.6	-36.0
16	GAAAGGGCTGGCTATGATAAACTGTGGC	28	0.50	-19.4	-33.7

A bases. Our double filtering approach accounts for both situations.

3.3 Methods

For validation purposes, we focus on the ENCODE regions of the human genome [50]. These 44 regions together comprise 1% of the human genome. The ENCODE regions were chosen to be representative of the entire genome, based on gene density, GC content, and density of conserved non-coding elements.

In addition, we chose a collection of sixteen query sequences to use in our experiments. We manually selected from the ENCODE regions exonic and intronic sequences that vary in length from 22 to 31 bases. Each selected sequence was analyzed using HYBRID, assuming standard PCR conditions (see below). A selected sequence S was added to the query set if the binding affinity between S and its reverse complement is greater than -19 kilocalories per mole. None of the selected query sequences overlaps a repeat sequence as annotated by RepeatMasker, and the percent GC of the queries range from 40% to 80%. The final list of query sequences is given in Table 3.3.

To generate a gold standard set of binding sites, we used HYBRID to scan every base of both strands of the ENCODE regions. The scan employed a window size of 35 bases, and was repeated with two different sets of experimental conditions, typical of PCRs and microarray experiments, respectively. For PCR conditions, we predict binding affinities at 55 C, with a concentration of 50 millimolar NaCl and 2 millimolar $MgCl_2$. For microarray conditions, we predict binding affinities at 40 C, with a concentration of 1 molar NaCl. In subsequent experiments, we used these lists of binding sites to verify that our algorithm correctly identifies all binding sites.

In selecting filter thresholds, we focus on two levels of binding site stringency, corresponding to weak and medium binding. We define a weak binding site as one where the equilibrium constant of the dimer formed by the binding site and the query sequence is at most six orders of magnitude less than the dimer formed by the query sequence binding to its reverse complement, under equal initial single strand concentrations. We define medium binding sites similarly, except we require only three orders of magnitude of difference. We used all binding sites recovered with the linear scans to choose filter thresholds.

3.4 Results

In order to measure the efficiency and accuracy of our binding site prediction algorithm, we scanned the ENCODE regions with a collection of query sequences, using HYBRID with and without the filtering and indexing pipeline.

We began by examining the behavior of each of the four filters for the thresholds designed to detect k-mers with high thermodynamic affinity to the query. Table 3.2 lists the percent of k-mers eliminated by the combined filters for each of the 16 query sequences. The mismatch kernel appeared to provide the most value, since it had a

Table 3.2: Rejection rates for the four filters. The table lists, for each of the query sequences in Table 3.3, the percentage of k-mers rejected by each of the four filters using the high affinity filter thresholds, as well as the total number of k-mers that pass through all four filters. These results are for weak binding sites in standard PCR conditions.

Sequence	F_1	F_2	F_3	F_4	Remaining
1	99.4	74.3	3.8	0.6	1589
2	99.2	69.0	7.3	2.8	2431
3	99.4	63.8	6.4	10.0	1862
4	99.3	54.0	2.9	24.0	2333
5	99.1	63.0	1.1	48.0	1887
6	99.3	51.7	3.1	29.7	2282
7	99.3	63.8	5.1	2.5	2322
8	99.2	65.0	7.3	14.5	2414
9	99.3	66.5	6.9	10.0	2059
10	99.5	58.5	2.9	13.6	1969
11	99.3	59.8	6.5	26.6	2164
12	99.3	63.2	5.7	47.1	1360
13	99.3	63.7	3.9	20.2	2181
14	99.4	75.4	2.8	2.7	1419
15	99.1	67.9	8.0	12.1	2375
16	99.2	68.5	4.0	47.4	1316
Mean	99.3	64.3	4.9	19.5	1998

rejection rate over 99%; however, this high rejection rate is primarily a result of its placement first in the filter pipeline. In practice, the more computationally expensive filters are also more exclusive. In each case, the filters reduced the complete set of $4^{10} = 1,048,576$ k-mers to less than 2500 k-mers per query. Also, note that the setup in Table 3.2 (weak binding sites in PCR conditions) is the most permissive and hence yielded a relatively large number of k-mers. Table 3.3 lists the total number of k-mers that successfully passed through all four filters in each experiment: medium and weak binding, and PCR and microarray conditions. With the exception of the weak binding/PCR conditions, the algorithm typically produced on the order of 20 k-mers for further consideration. The results in Tables 3.2 and 3.3 used the filter thresholds

Table 3.3: k-mer filtering performance. The table lists, for each of the query sequences in Table 3.3, the total number of k-mers that passed through all four filters using the high affinity thresholds.

Sequence	PCR Weak	PCR Medium	Microarray Weak	Microarray Medium
1	1589	15	15	15
2	2431	20	20	20
3	1862	14	14	14
4	2333	16	16	16
5	1887	19	20	16
6	2282	16	16	16
7	2322	16	16	16
8	2414	20	20	20
9	2059	17	17	17
10	1969	13	13	13
11	2164	18	18	18
12	1360	15	17	14
13	2181	18	18	18
14	1419	14	14	14
15	2375	21	21	21
16	1316	16	17	13
mean	1997.7	16.8	17.0	16.3

selected using the high affinity filter parameters. Results for the high similarity set of thresholds were similar.

After obtaining both groups of candidate binding site anchors, we then identified locations in the sequence where a k-mer from the high affinity group occurs near a k-mer from the high similarity group. Table 3.4 lists, for all four experiments, the percentage of sites identified by the high affinity group of binding site anchor candidates that were not close enough to a k-mer occurrence from the high similarity group of binding site anchor candidates. On average, this step reduces the list of candidate sites by between 63% and 85%, depending upon the experiment.

The final stage of the analysis involved running HYBRID on the filtered list of

Table 3.4: Proximity filtering performance. The table lists, for each of the query sequences in Table 3.3, the percentage of sequence locations that were rejected by the proximity filtering step. The final row contains the column average.

Sequence	PCR Weak	PCR Medium	Microarray Weak	Microarray Medium
1	69	88	66	94
2	55	71	50	57
3	59	77	71	83
4	69	80	70	94
5	76	93	91	92
6	57	67	63	74
7	57	48	48	64
8	80	98	81	98
9	54	70	62	82
10	58	72	59	91
11	61	65	64	73
12	63	74	76	85
13	68	94	80	97
14	48	69	56	89
15	74	89	79	95
16	63	85	79	88
mean	63.19	77.50	68.44	84.75

candidate binding sites. Table 3.5 lists, for each experiment, the number of candidate binding sites that were evaluated by the HYBRID software. Clearly, this stage is very important, since the number of sites considered is typically several orders of magnitude larger than the number of sites that HYBRID identifies as binding partners. In this sense, our filters are conservative: they do not very closely approximate the computation performed by HYBRID. However, these conservative thresholds lead to high accuracy. For all 16 primers that we tested, our filtering and indexing pipeline identifies 100% of the binding sites that were identified by HYBRID in the much more computationally expensive linear scan of the entire ENCODE regions. Furthermore, as shown in Table 3.5, the entire pipeline is very efficient. For medium binding

Table 3.5: Number of candidate sequences examined and accepted by the partition function model of DNA binding, and time for each run. The table lists, for each experiment, the total number of candidate sites produced by the filtering and indexing pipeline, the number of those sites that are considered by HYBRID to be true binding sites, and the total wall clock time required to identify the sites.

Sequence	weak PCR Cand.	weak PCR Actual	weak PCR Time	medium PCR Cand.	medium PCR Actual	medium PCR Time	weak microarray Cand.	weak microarray Actual	weak microarray Time	medium microarray Cand.	medium microarray Actual	medium microarray Time
1	30712	25	6 m	2340	19	18 s	9543	21	35 s	1332	16	23 s
2	57587	23	20 m	11994	15	40 s	18702	16	76 s	3326	11	19 s
3	44628	100	8 m	4030	20	21 s	4882	21	17 s	3078	17	19 s
4	35218	45	11 m	5269	19	22 s	6152	20	24 s	1178	16	12 s
5	23235	29	6 m	1870	13	18 s	2791	14	23 s	1132	9	12 s
6	91220	108	13 m	7304	19	26 s	8112	21	28 s	6301	16	21 s
7	33780	48	10 m	6667	20	31 s	6667	22	24 s	4962	15	28 s
8	22310	26	5 m	179	14	10 s	3025	15	19 s	99	10	12 s
9	35396	45	12 m	6552	18	22 s	8390	19	35 s	4264	14	17 s
10	175109	336	12 m	5741	21	17 s	7976	25	29 s	1909	18	11 s
11	75547	40	16 m	5908	17	24 s	6181	18	32 s	4896	14	28 s
12	20887	70	6 m	3120	18	16 s	3369	20	19 s	1598	14	14 s
13	22934	31	6 m	907	19	14 s	3276	20	16 s	418	14	13 s
14	142717	361	13 m	5221	21	22 s	7081	37	34 s	1837	17	13 s
15	20106	26	8 m	1413	14	13 s	2839	16	17 s	153	10	12 s
16	17138	29	6 m	2988	17	17 s	3680	19	19 s	1639	13	17 s
mean	53032.8	83.9	9.9 m	4468.9	17.8	20.7 s	6416.6	20.3	27.9 s	2382.6	14.0	16.9 s

strength and standard PCR conditions, HYBRID was only required to evaluate an average of 4469 sites, and scanning the entire ENCODE database required 20.7 seconds on average. By comparison, a linear scan of the ENCODE regions using HYBRID takes approximately two days.

Figure 3.5 shows an example of a query hybridizing to a weak binding site. Here, the free energy of binding is -10.9 kcal/mole, sufficient so that a substantial fraction of sites would be occupied by a primer at thermodynamic equilibrium. Figure 3.6 shows an example of a query hybridizing to a medium binding site. Here, in spite of another internal mismatch, the overall free energy of binding is much higher at -16 kcal/mole, and would result in binding of nearly all copies of the template by the primer under typical PCR conditions. The medium example illustrates the sequence specificity of the stability of internal mismatches: the free energy of the GG mismatch is nearly -2 kcal/mole, and hence stabilizing, whereas the CT mismatch contributes 2 kcal/mole, and is destablizing to the helix.

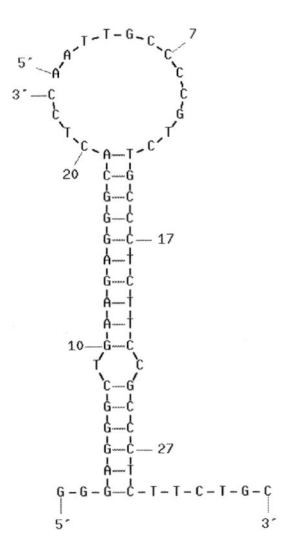

Figure 3.5: Example of weak binding site. ΔG is -10.9 kcal/mole

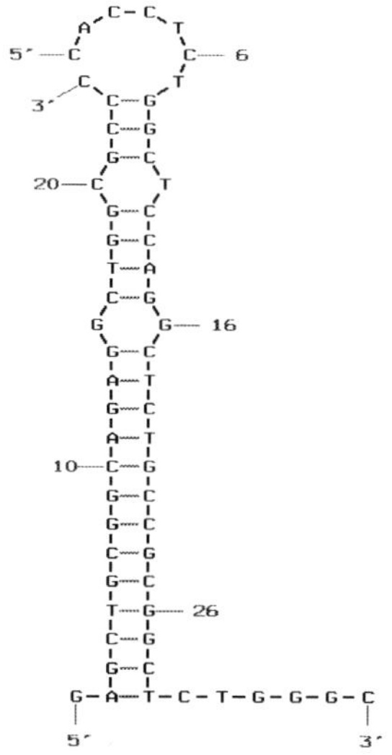

47

```
           C- C
         A´     `T
    5´—C       C—— 6
       `C     T´
      3´ `C—G´
          C---G
          C—C
          G—C
    20——C     T
         `G—C´
          G---C
          T---A
          C----G
         G     G——16
         `G—C´
          A—T
          G—C
          A—T
    10——C—G
          G—C
          G—C
          C—G
          G—C
          T---G——26
          C---G
          G—C
      G—A—T—C—T—G—G—G—C
      |                |
      5´               3´
```

Figure 3.6: Example of medium site. ΔG is -16 kcal/mole.

3.5 Discussion

We have presented a method for rapidly identifying binding partners for a given query DNA sequence within a genome-sized DNA database. Our approach combines a k-mer filtering method, which identifies k-mers that could nucleate binding sites to the query, with an efficient indexing method, which rapidly locates these nucleating k-mers in a sequence database. The combination of these two methods speeds up the DNA binding site search by at least three orders of magnitude.

We note that not all predicted binding sites will be relevant to every hybridization reaction. Some dimers may be slow to reach equilibrium concentrations, especially if the dimer has internal loops. Thus, in a PCR, some dimers may not have time to form and thus may not be a problem. However, in microarray hybridization experiments, conditions are much closer to equilibrium, and secondary binding sites may be more of a concern.

Among the four tasks that we considered, finding weak binding partners for PCR primers is the most difficult search task, and the one for which we obtain the least improvement. However, this task may be the most important for experimentalists, because even weak binding sites can drive high yields on undesired background reactions. This is because in PCR, the primers are present in vast excess, and the excess concentration of primer in the initial stages of the reaction drives high levels of weak binding site occupancy, even though the binding affinity is low.

The major bottleneck in our method is evaluating the final list of sequences. Even though we reduce the number of sequences that must be considered by several orders of magnitude, the partition function model is still sufficiently slow that it introduces a significant computational burden. It is important to recognize, however, that we can typically place an upper limit on this burden: once we identify a pre-specified number of binding partners for a given query, the search can terminate, since that particular query is not a tenable primer or probe candidate.

3.6 Future Work

Conceptually, searching an RNA database for binding sites to an RNA sequence is similar to the problem addressed in this paper. Although the same partition function model can be used to compute the binding affinity of one RNA molecule for another, the parameters are different due to the chemical differences between RNA and DNA [113]. We are currently beginning experiments to evaluate the computational complexity of this version of the binding site search problem. Further, it may also be of interest to search for DNA binding partners of an RNA molecule, or RNA binding partners for a DNA molecule. Because the data for these heterogeneous dimers is much less complete than the data for DNA/DNA or RNA/RNA dimers, our method is not applicable to these binding site searches.

Our method depends critically on the filter parameters, and clearly the similarity of the anchoring k-mers in a binding site to a query is not known in advance. We are therefore increasing the size of our database of predicted binding sites, so that we can estimate the sensitivity of our method for a wider variety of query sequences.

3.7 Conclusions

We have shown that DNA binding site search of genomic scale DNA sequences is tractable for realistic experimental conditions, for primer length DNA sequences. Our filters work together to reduce by at least three orders of magnitude the number of sequences that must be examined by a partition function model of DNA binding, reducing search time from two days to scan the ENCODE regions to under a minute for typical queries. This filter- and index-based method will be useful in the design of PCR primers and short oligonucleotide probes.

Chapter 4

DNA THERMODYNAMICS APPLIED TO PCR PRIMER DESIGN

Introduction

The polymerase chain reaction (PCR) [170], a method for making many copies of a specific DNA fragment, is one of the most widely applied tools in modern molecular biology [82]. PCR applications range from directed mutagenesis to disease diagnosis. Crucial to the success of a PCR is the choice of the short oligonucleotide sequences (primers) that flank the fragment to be copied (the template). These primers must fulfill a number of criteria, and research into primer selection has been ongoing since the advent of PCR [167, 106, 73, 164, 172]. Primer design is an unsolved problem, especially in genomics approaches where regions must be comprehensively analyzed by PCR assays. We focus especially on PCR primer design for regions in repeated sequences, because repeated sequences are not amenable to standard primer design approaches and yet comprise a significant fraction of mammalian genomes.

Our motivation is to develop physically guided methods for predicting primer quality in order to improve primer design. The primary difficulty that we seek to address in PCR primer design is how to predict primer quality, defined here as the ability to efficiently and specifically amplify the desired template fragment, on the basis of the primer sequences and the template. Our physically motivated methods have two significant benefits compared to ad hoc primer scoring schemes. First, they take advantage of accurate methods for assessing DNA binding [60, 45] and folding stability [115]; these accurate assessments are critical because PCR relies fundamentally on DNA binding reactions. Second, a physically motivated approach reduces the number of parameters that must be chosen, and shifts the emphasis from choosing arbitrary thresholds for quality scoring metrics to specifying physically meaningful reaction conditions and primer quality criteria.

Early methods for primer design proposed a variety of quality metrics in order to evaluate various aspects of primer quality, and then combined these individual metrics into a final metric using a weighted sum [167, 106, 73]. These quality metrics were based on considerations such as primer melting temperature, the thermodynamic stability of the primer at the 3' end, and a variety of other factors motivated by practical experience with PCR. Because many metrics contribute to the final prediction of primer quality, a weight for each individual quality metric must be specified in order to obtain the final primer pair score. The user of these methods must select these weights appropriately for the PCR application at hand.

However, selecting these quality metric weights presents two significant difficulties. First, these metrics are not always physically accurate, and second, they can be redundant. For example, good PCR primers should not stably bind to other primers (forming so called primer dimers); if they bind stably to other primers, then they are much less likely to participate in the desired priming reaction. The widely used program Primer3 [164] uses two metrics to assess the likelihood of a primer binding to itself or another other primer: the max-complementarity metric and the max-3'-complementarity metric. The difference between these two metrics is that one considers overall similarity between two primer sequences, and the other considers similarity anchored at the 3' ends, as computed by Smith-Waterman alignment scores. These metrics are redundant because high 3' anchored similarity implies high overall similarity, and these metrics are thermodynamically inaccurate because they do not account for known effects in DNA binding interactions such as the sequence specificity of single internal mismatches [83]. Consequently, selecting appropriate weights for these two metrics for the final quality evaluation presents significant difficulties. These difficulties are compounded by the large number of quality metrics that must be weighted for the final primer quality metric.

We developed a thermodynamic approach for assessing primer pair quality that uses statistical mechanical models of nucleic acid binding and folding [60, 45, 115], coupled with chemical reaction equilibrium analysis [185]. We compute binding and folding affinities for a variety of relevant dimer and folded forms and then solve for the concentrations of all species at equilibrium. We then identify a primer pair as acceptable if the desired chemical species (primers annealed to priming sites) are present at high concentration at thermodynamic equilibrium. Although PCR is not an equilibrium process, our equilibrium analysis is essentially a conservative method of identifying problematic primers that have a propensity for folding or undesired dimerization behavior.

Results

Algorithm Development

We address the problem of choosing acceptable and specific PCR primers for a locus given a genomic DNA sequence, a set of user supplied parameters and constraints, and the coordinates of the locus. Our primer design program calculates binding and folding energies for a variety of relevant chemical species, and then integrates these calculations into a final measure of PCR efficiency. Below, we describe how these energies are computed and then integrated into our final quality metric. Because computing the final primer efficiency measure is a bottleneck in the screening of primer candidates, we then describe a machine learning approach to predict primer acceptability on the basis of free energy calculations; this classification approach allows us to quickly eliminate infeasible candidates.

In addition to predicting whether the primers will amplify a given locus, we also evaluate the primer specificity. Specific primers will amplify only the desired locus, whereas non-specific primers have binding sites in the background DNA that lead to undesired copying of background fragments in addition to the target locus. In order to predict primer specificity, we use a precomputed index, in conjunction with a thermodynamic heuristic for predicting primer specificity. Following **(author?)** [124], we identify the shortest sequence at the 3' end of each primer that could bind stably and then identify exact occurrences of this sequence in the background genomic DNA using our precomputed index. Although this approach will miss some binding sites, we show that it works well in practice.

In order to test our approach, we compared our method to a highly optimized primer selection strategy used for several high-throughput studies [48, 168, 169]. This method used Primer3 for designing primers and a method focused on the 16 bases at the 3' end of each primer for predicting specificity. We focused on the problem of tiling genomic regions, in which primers are placed to cover as much of a selected genomic region as possible, with minimal overlap between adjacent PCR products. We show in this work that our approach to evaluating primer quality and specificity is more accurate than current approaches. Furthermore, our approach has fewer adjustable parameters than current approaches, and these parameters are more physically meaningful. Thus, our approach is easier to tailor to specific reaction requirements.

Our method, Pythia, takes as input the genomic sequence, locus coordinates to be amplified, and user specified parameters. Figure 4.1 illustrates our method. In step 1, Pythia identifies all pairs of sequences that satisfy the user constraints, such as primer melting temperature, primer length, and amplicon length. Pythia then sorts these primers by the discrepancy between the desired melting temperature and the average of the computed primer melting temperatures. Pythia then examines the candidates on the list. In step 2, Pythia computes the quality metric for the primer pair, and if that metric is above a user specified threshold, then Pythia computes a specificity check as step 3. If the primers meet the specificity criterion, then Pythia outputs the primer pair. If the equilibrium efficiency is not above the user specified threshold or the primers are not specific, Pythia examines the next candidate. Pythia proceeds in

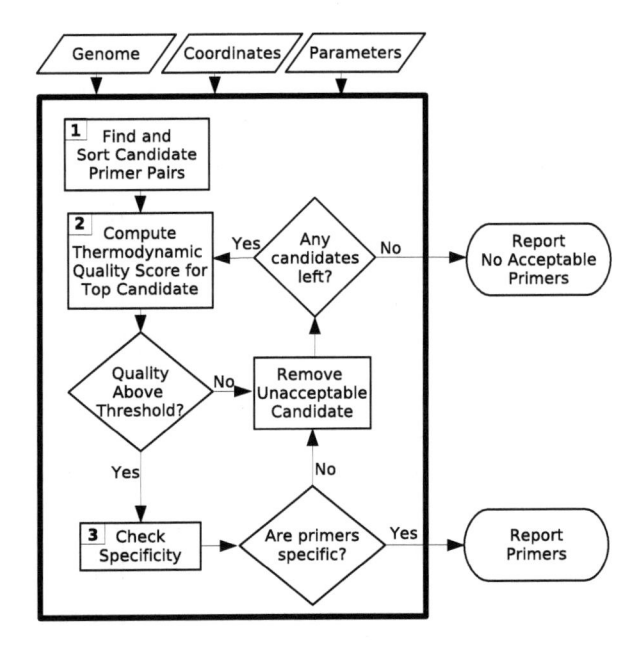

Figure 4.1: Flowchart of the Pythia algorithm. Inputs are the genomic sequence, locus coordinates, and user specified parameters. In step 1, Pythia identifies all primer pairs meeting the user specified requirements and sorts these primer pairs by the sum of the differences between the computed and target primer melting temperatures. In step 2, Pythia computes the thermodynamic quality metric for the top ranked candidate. If this candidate meets a user specified metric threshold, then Pythia proceeds to step 3. If not, the top ranked candidate is removed from the list and Pythia returns to step 2. In step 3, Pythia performs a specificity check. If the primer passes the specificity check, it is given to the user, and the program terminates. If not, the top ranked candidate is removed from the list and Pythia returns to step 2.

this way until a feasible candidate is found, or until no candidates are left.

In order to predict the binding affinity of two nucleic acid molecules, we use reference data [174] that describes the thermodynamic stability of DNA base pairing and stacking interactions, as well as the stability contributions of loops (unbound bases between helical segments). We use a statistical mechanical method [45] in conjunction with the reference data to predict the overall binding affinity.

Reference data used to predict the stability of DNA binding interactions has been steadily refined [22, 171, 174] and now includes stability parameters for an assortment of features such as base pairing and stacking, hairpins, and internal loops. Using this reference data, the melting temperature of a DNA duplex can be accurately predicted using simple rules [174] when the two sequences are complementary or include isolated internal mismatches.

When the two sequences in a duplex are not complementary, predicting their binding affinity requires more sophisticated approaches. We use statistical mechanical models of DNA to compute the binding affinity between the relevant DNA molecules in a PCR reaction [45, 60]. These models use dynamic programming to evaluate the stability of many configurations in which one molecule is bound to the other via at least one base pair and integrate the stabilities of all of these conformations into a final stability prediction.

In addition to primer binding, we account for primer folding, because primers with stable folded structures may not be available for the desired priming reactions. A different statistical mechanical approach has been developed to predict the folding energy of a nucleic acid molecule [115]. We do not consider folding conformations with pseudoknots, so that we can employ a dynamic programming algorithm with a computational complexity of $O(n^4)$ in the length of the folded sequence rather than $O(n^7)$ [46] when pseudo-knots are considered.

We use these free energies to assess the stabilities of dimer and folded forms. However, because many reactions are competing for DNA strands in PCR, we use

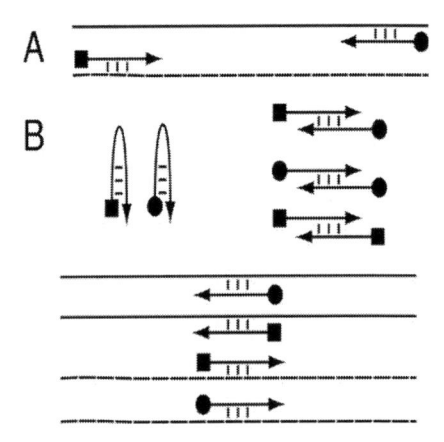

Figure 4.2: Species accounted for in primer feasibility analysis. The solid line is the top strand of the template; the dashed line is the bottom strand of the template; the arrow with the square end is the left primer; the arrow with the round end is the right primer; three dashed lines indicate binding (or folding) via hydrogen bonding. (A) Desired binding interactions. High rates of binding are desired between the primers and the template priming regions. (B) Undesired binding and folding reactions. Primers should not fold, dimerize, or bind to the target outside of the priming regions.

coupled equilibrium analysis to assess the equilibrium concentrations of the relevant species.

Equilibrium analysis

In a PCR, many reactions are simultaneously competing for single unbound target fragments. We consider 11 reactions that compete for single unbound strands; these reactions are depicted in Figure 4.2. In particular, we consider primer folding, primer dimerization, primers binding to template outside of the priming region, and primers binding to template in the priming region. Of these reactions, only the last type is desired; the rest should be minimized. However, PCR can work in the presence of some primer folding and dimerization, provided the primers bind well to the priming regions. In order to balance these considerations, we use chemical reaction equilibrium analysis [185].

Chemical reaction equilibrium analysis determines the concentration of each chemical species at thermodynamic equilibrium; in this context, we obtain the concentration of each DNA folded, unfolded, and dimer species. In order to evaluate the feasibility of a primer pair, we compute the free energy of all of the duplex and folded forms and then compute the equilibrium concentration of all of these species as described in the supplement. In order to characterize the quality of the primer pair, we use a quantity that characterizes the efficiency of PCR assuming equilibrium binding conditions. In particular, we determine the equilibrium efficiency as the minimum of the fraction of left primers binding to the left primer binding site and the fraction of the right primers binding to the right primer binding site. We choose the minimum of these fractions because a PCR can only be as efficient as its least efficient priming reaction.

Of course, PCR is manifestly not an equilibrium reaction. Our use of equilibrium analysis is designed to detect potential problems by identifying binding and folding reactions that are significant enough to disrupt priming. We assume that if a primer pair works under our equilibrium model, then it will work in PCR conditions. The converse is not true; because some dimerization reactions may be kinetically slow, some binding interactions that are problematic at thermodynamic equilibrium may not be relevant under PCR conditions. Nevertheless, our approach rejects primer pairs in which equilibrium binding conditions result in insufficient binding of primers to their priming sites in the template molecules.

Support Vector Machine Prediction of Feasibility

In typical primer design problems, on the order of ten thousand primer pairs satisfy the user supplied constraints (such as melting temperature and length restrictions).

Because the gradient descent procedure requires many relatively slow $O(n^3)$ matrix inversion steps for each update to the solution, we developed a filtering procedure to quickly reject infeasible candidates.

Our approach is to use a support vector machine classifier [40] to predict whether a primer pair would meet an efficiency threshold if the full equilibrium analysis were run, on the basis of the free energies of the various species we consider. A support vector machine uses a hyperplane to classify a sample on the basis of a vector of features in a feature space. Support vector machines are widely used in computational biology [134] and have been applied to many bioinformatics problems such as translation site initiation recognition [223], microarray analysis [23], and genome annotation [153].

A critical component of a support vector machine classifier is the design of feature vectors associated with the samples. We designed our feature vectors to account for the intuition that in a system with many competing reactions, it is not the absolute free energy of any particular reaction that is important, but rather the relative free energy of a reaction as compared to its competitors. We therefore used a quadratic kernel [40] on vectors consisting of the 11 free energy values that we compute for each primer pair; this quadratic kernel provides information on all pairs of free energy values to the classifier. For further speed improvement, we explicitly compute the weight vector so that we can compute the classifier decision function as an inner product rather than a kernel expansion. We trained the support vector machine using LibSVM [32].

Primer Design Algorithm

Our primer design algorithm requires the user to specify a set of constraints, annealing conditions, and quality threshold. The constraints are the allowable range of melting temperatures, primer lengths, and amplicon lengths. The user must also specify the PCR annealing temperature, the primer and template concentrations, and the salt concentration. Finally the user must specify the minimum equilibrium efficiency for the primer pair quality metric, and the stability threshold and maximum non-specific product length for the specificity check. For one of the constraint sets, the minimum and maximum primer melting temperature, the user also selects a target melting temperature. Given a set of constraints, all primer pairs for a locus satisfying the constraints are enumerated and then sorted by the sum of the differences between the target melting temperature and the melting temperature for each primer. We sort initially by the primer melting temperature differences so that the resulting primers have acceptable equilibrium behavior and also uniform thermodynamic characteristics.

For each element of the sorted list, the 11 free energy values are computed and screened by the support vector machine classifier. If the classifier flags the primer pair as feasible, then the full equilibrium analysis is performed and the quality metric for the primer pair is computed. If the quality metric exceeds a user defined threshold, then the specificity analysis is run. The specificity analysis identifies the shortest region of the 3' ends of each primer that exceeds a stability threshold, and then identifies exact occurrences of these sequences in the background genomic sequence (see supplement). If any of these hits are close and oriented appropriately to generate undesired PCR amplicons, then the primers are flagged as non-specific; otherwise they are flagged as specific. The first primer pair that exceeds the quality metric and which is flagged as specific is returned.

Comparison to Other Methods

In order to evaluate our approach, we compare it to a highly optimized primer selection strategy used for several high-throughput studies [48, 168, 169]. This approach uses carefully chosen parameters for Primer3 and a method for assessing primer specificity based on the 16 bases at the primer 3' end. In this approach, exact occurrences of the sequence formed by the 16 bases at the 3' end of each candidate primer are located in the genome, and if there are too many occurences of either sequence, the primer pair is rejected. We refer to this combination of Primer3 and the 3' end-based specificity evaluation as P316.

We first evaluated our approach before developing the support vector machine classifier to predict primer feasibility based on free energies. For this test, we selected three regions of the human genome for which tiling primers had already been designed by the P316 method. Because computing the solution to our coupled equilibrium problem requires about 0.7 seconds of computation, and a typical region has on the order of 10,000 primer candidates (100 candidates for the left primer and 100 for the right), we limited the amount of time our program was allowed to attempt to design primers for any particular interval to 10 minutes, thus allowing Pythia to consider at most approximately 900 candidates per interval.

Motivated by the bottleneck induced by the coupled equilibrium analysis, we then developed the support vector machine classifier, which was fast enough so that Pythia could evaluate all of the candidates in a region if necessary. We then chose to tile short regions near transcription start sites annotated as interspersed repeats, because these regions were challenging for the methods employed by the P316 approach.

We evaluate each method by the fraction of successful PCRs. Because we use melting curve analysis to assess each PCR, we must infer the success rates of each method and the coverage based on the success rates of a selected group of PCRs that were analyzed both by melting curve analysis and by running the PCR products on an agarose gel.

Validation by Melting Curve Analysis

We chose a set of PCRs not used in the primer design comparison to run on a gel in order to evaluate the melting curve analysis of PCR success. In melting curve analysis, the reaction mixture is slowly heated after thermal cycling to a temperature high enough to denature the PCR amplicons. Because amplicon denaturation typically occurs in a narrow temperature interval [146, 186], the fluorescence used in qPCR to detect double stranded DNA will decrease sharply in the temperature range in which

Table 4.1: Concordance between agarose gel and PCR amplicon melting curve results. For a selected set of PCR primers, we compared the results of melting curve analysis to agarose gel analysis of PCR amplicon. Melting curves were classified as valid or invalid based on melting curve morphology, and gel lanes were classified as clean or not clean at two levels of stringency. In each table entry, the numbers correspond to the number of reactions with the corresponding gel and melting curve label at stringent and permissive levels of gel scoring stringency.

Gel Label	Valid Melting Curve		Invalid Melting Curve	
	Stringent	Permissive	Stringent	Permissive
Clean	172	199	33	41
Not Clean	38	11	16	8

the PCR amplicon denatures. A plot of the negative first derivative of this fluorescence will yield a single prominent peak for PCRs in which the amplicon molecules denature in a narrow range of temperatures. Melting curves were scored manually as valid if they had a single prominent peak, and invalid if they had multiple prominent peaks or other unusual morphology.

We ran 259 PCR products on agarose gels stained with the dye Sybr Green I. We manually examined the lanes and marked them as clean or not according to two levels of stringency. Under a permissive scoring system, lanes were marked as not clean if there was significant smearing, missing bands, or prominent additional bands in addition to the band of the expected size. Under a stringent scoring system, all bands marked not clean under the permissive system were also marked not clean, as well as all bands with faint additional bands or faint smearing.

We used melting curve analysis to verify synthesis of a PCR product in each reaction, and hoped to use it to verify PCR specificity. However, analysis of the concordance between gel and melting curve labels shows that melting curve morphology only weakly correlates with gel results. Table 4.1 shows the concordance between the gel labels and the melting curve labels for both levels of stringency. This result shows that although the valid melting curves are enriched for clean gel lanes, the in-

Table 4.2: Genomic characteristics of selected human genome regions. We compared the ability of Pythia to the ability of the P316 algorithm to tile these regions. We show the location, size, and a brief description of each locus.

Region	Chromo-some	Interval Start	Interval Stop	Length	Description
1	16	147,000	164,000	17Kb	High GC Content
2	16	181,000	215,000	34Kb	Repetitive
3	11	5,252,000	5,277,000	25Kb	Typical

valid melting curves also mostly correspond to clean gel lanes. Thus, melting curves have limited utility in discriminating aberrant PCRs from clean PCRs at both levels of stringency in gel labeling. For the permissive scoring system, 95 percent of valid melting curves had a clean lane, and 84 percent of the invalid melting curves were also clean. For the stringent scoring system, 82 percent of valid melting curves were clean, compared to 67 percent of invalid melting curves. Four lanes did not have bands of the expected size, and all corresponded to invalid melting curves.

Based on this result, we compute the success rates by extrapolating from the stringent success rates and the permissive success rates (see Methods for our extrapolation).

Application to Genomic Tiling

We chose three regions for which primers had already been designed for the first evaluation of our approach. Table 4.2 summarizes the three regions that we tiled in the first test of our method. We attempted to tile these regions as densely as possible with PCR products whose size ranged from 225 to 275 bases, and whose primers had melting temperatures ranging from 60C to 64C, with a target of 62C. Primers were constrained in length to lie between 18 and 30 bases.

We attempted to design a PCR primer for the first 275 base pair window in the

Table 4.3: Primer design performance for selected human regions. Shown are the number of PCRs and the extrapolated success rates for permissive and stringent criteria.

Region	P316 PCRs	P316 Success Rate Permissive	P316 Success Rate Stringent	Pythia PCRs	Pythia Success Rate Permissive	Pythia Success Rate Stringent
1	49	94%	80%	41	94%	81%
2	93	94%	81%	102	94%	81%
3	63	92%	78%	43	94%	81%

region. Pythia was allowed to spend at most ten minutes on each interval. If Pythia was able to choose a primer pair in the allotted time, we then attempted to design primers for the 275 base pair window starting at the end of the last successful design. If Pythia was not able to design primers for the window, then we moved the window by 25 bases and tried again. We stopped this iterative process when the design window reached the end of the region. We then attempted to fill gaps by attempting to tile the gaps, increasing the time allowed per interval to 20 minutes.

Even when constrained in the time allowed to design primers, Pythia achieves comparable performance to P316 on human genomic intervals. Table 4.3 shows that Pythia achieves comparable success rates and attempts to place slightly fewer primers in two of the three regions. Examination of this data revealed that for some regions, Pythia must consider on the order of ten thousand primer pairs. However, due to the time required for equilibrium analysis, Pythia could only evaluate approximately 900 candidates in the 10 allotted minutes. In order to increase the number of candidates that Pythia could examine in a fixed amount of time, we developed an SVM approach to screening primer candidates.

Evaluation of Primer Feasibility Classifier

We next evaluated the accuracy of our primer feasibility classifier. First, we collected candidate primer pair examples for seven human genomic loci, and computed the equilibrium efficiency metric for each example. We then trained a support vector machine to predict whether the equilibrium efficiency was above a threshold for several threshold choices and evaluated classifier performance using five-fold cross validation. For each choice of threshold, we selected all of the negative examples and an equal number of positive examples. Support vector machines require a parameter to specify the trade off between model accuracy and complexity; we set this cost parameter to 0.1.

We used receiver operating characteristic (ROC) analysis [120] to evaluate the performance of our classifier. An ROC curve plots the true positive fraction against the false positive fraction for a range of decision function values. The area under this curve, the ROC score, is a measure of how well the classifier is able to distinguish between the two classes: an area of 0.5 is the expected area under the ROC curve for a random classifier, and an area of 1.0 is the area under the ROC curve for a perfectly accurate classifier.

We used five-fold cross validation to evaluate the ability of the SVM to predict the results of equilibrium analysis on data which was not used in training. We split each dataset randomly into five parts, and trained the classifer on data from four of the parts. We evaluated its performance using ROC analysis on the fifth part. For our final classifier evaluation, we computed the average ROC score over all five portions of the data.

Our results show that the classifiers are able to learn to distinguish between acceptable primer pairs and unacceptable primer pairs with high accuracy, and thus predict, given a set of free energies, whether the minimum equilibrium binding fractions are above the specified thresholds. Table 4.4 shows the training set sizes and the

Table 4.4: The number of training points for each acceptability threshold. For each threshold, we show the number of examples used to train the SVM, and the ROC and ROC50 scores. We assessed SVM performance using 5-fold cross validation.

Threshold	Dataset Size	ROC Score
0.8	642	0.9995
0.85	1474	0.9986
0.9	3056	0.9951
0.95	10498	0.9937

mean ROC score over all cross validation folds. For each choice of threshold, the ROC scores were above 0.99. Thus, the classifier can accurately filter primer candidates at low computational cost.

The computational savings are due to the nature of the rule that the support vector machine uses to classify data. This rule associates a weight with each of the input features, and the classifier decision is made by computing the sum of the input features multiplied by the corresponding weights. If this sum is greater than zero, then the SVM classifies a datapoint as acceptable according to equilibrium analysis, and unacceptable otherwise. Because we use a quadratic kernel on a vector with 11 features, we can screen primers pairs on the basis of the free energies with just 264 multiplications and 131 additions by explicitly using the weight vector; this is a substantial efficiency improvement over applying the equilibrium analysis to each primer candidate.

Application to Repetitive Elements

After developing the SVM classifier to screen primer candidates, we applied our method to tile a set of regions near transcription start sites that were annotated as interspersed repeats by the RepeatMasker program [184]. We designed primers to tile each region along with 125 bases flanking each end. Because the PCR products

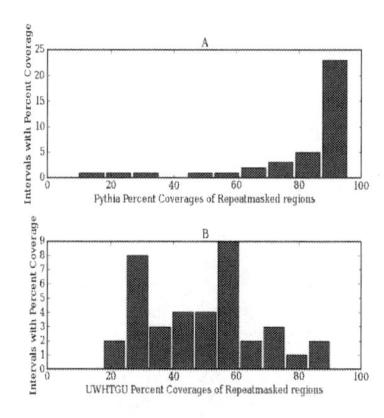

Figure 4.3: Primer pair design coverages for interspersed repeat regions. Design coverage is defined as the fraction of an interval covered by PCR product sequences. (A) Histogram of coverages for Pythia (mean 80%). (B) Histogram of coverages for P316 (mean 50%).

were between 225 and 275 bases in length, each primer pair had at least one primer in a repeat-annotated region. We designed primers to tile 38 such intervals with a mean length of 1.5kb (where the minimum interval length was 751 bases and the maximum interval length was 6198 bases).

For these regions, Pythia was able to design primers for much greater coverage. Figure 4.3 shows a histogram of the percentages of each region that were covered by primer pairs designed by Pythia or the P316 approach. Pythia designed 195 primer pairs to tile these regions, whereas the P316 method designed 106 primer pairs to tile these regions. Based on melting curve analysis, Pythia achieved a 94% success rate under the permissive criteria and an 80% success rate under the stringent criteria; similarly, the P316 approach achieved a 95% success rate under the permissive criteria and an 82% success rate under the stringent criteria. Of the 38 regions, Pythia was able to design primers to cover at least 80% of 27 regions, whereas the P316 approach

Table 4.5: Pythia acceptability assessment of P316 primers. We assessed the ability of the Pythia primer pair quality metric to predict the quality of the P316 primers. Here, we compared the Pythia primer assessment to the results of melting curve analysis.

Pythia Evaluation	Valid Melting Curve	Invalid Melting Curve
Acceptable	276	15
Unacceptable	17	3

was able to design primers to cover at least 80% of only 2 regions. In contrast, Pythia was able to design primers to cover less than 50% of only 3 regions, compared with 18 regions with less than 50% coverage for the P316 approach.

Primer Quality Prediction

Crucial to the success of a primer design method is how well it can assess the quality of a primer pair. We therefore sought to compare the primer pair scoring functions in order to assess how well they can assess the likelihood that a primer pair will produce a product in a PCR. To assess the accuracy of these functions, we used Primer3 to assess the primers designed by Pythia, and we used the Pythia primer scoring function to assess the primers designed by the P316 approach.

The majority (94%) of the P316 primers were acceptable by the standards of the Pythia scoring function. Table 4.5 shows the results of Pythia analysis of the P316 primers. Pythia's primer design approach is conservative: most of the primers which Pythia scored as unacceptable (85%) resulted in acceptable amplicons.

Interestingly, the Primer3 primer metric rejected almost all of Pythia's primers. Table 4.6 shows the results of Primer3 analysis of the Pythia primers. About 95% of the Pythia primers were scored as unacceptable by the Primer3 scoring function, with only three of the unacceptable primer pairs resulting in failed PCR as judged by melting curve analysis. An informal examination of the Primer3 output revealed

Table 4.6: Primer3 acceptability assessment of Pythia primers. We assessed the ability of the Primer3 scoring function to predict the quality of the Pythia primers. Here we compare the P316 primer assessment to the results of melting curve analysis.

P316 Evaluation	Valid Melting Curve	Invalid Melting Curve
Acceptable	17	3
Unacceptable	322	39

Table 4.7: Pythia specificity assessment of P316 primers. We assessed the ability of the Pythia specificity assessment to predict which amplicons would yield non-specific products. Shown is the results of the Pythia specificity assessment of P316 primers compared to the labels of melting curves.

Pythia Evaluation	Valid Melting Curve	Invalid Melting Curve
Specific	280	29
Nonspecific	2	0

that no single property of Pythia's primers led to their rejection by Primer3. Rather, Pythia's primers collectively violated a variety of Primer3's primer evaluation rules.

Primer Specificity Assessment

Specificity is also a paramount concern in many contexts. Not only should a primer pair successfully amplify a target locus, but it should not amplify any other loci. We compared the specificity assessment methods for Pythia and the P316 primer design approach, to assess how well they could predict whether a primer pair will be specific.

Most of the P316 primers were predicted to be specific using Pythia's specificity heuristic. Table 4.7 shows the results of the Pythia specificity assessment applied to the P316 primer pairs. Pythia scored nearly all P316 primer pairs as specific (99%), and 29 of the primer pairs scored as specific had melting curves scored invalid. Only

Table 4.8: P316 specificity assessment of Pythia primers. We assessed the ability of the P316 specificity assessment to predict which amplicons would yield non-specific products. Shown is the results of the P316 specificity assessment of Pythia primers compared to the labels of melting curves.

P316 Evaluation	Valid Melting Curve	Invalid Melting Curve
Specific	17	9
Nonspecific	322	33

two primer pairs were scored as non-specific, and both of these yielded single peak melting curves.

In contrast, P316 predicted that most of Pythia's primers would be non-specific. Table 4.8 shows the results of the P316 specificity assessment of the Pythia primers. The P316 specificity assessment predicted that 93% of the Pythia primers would be non-specific. Of the primers with melting curves scored as invalid, 79% were scored as nonspecific, and of the primers with melting curves scored as valid, 95% were scored as non-specific.

Discussion

We propose our measure of equilibrium efficiency as a physically motivated criteria for predicting primer quality based on DNA thermodynamics. We have shown that our approach compares favorably to the P316 primer design approach, which is based on Primer3, and thus our approach has significant advantages when attempting PCR in RepeatMasked regions. Repeat sequences are important genomic features, comprising significant fractions of mammalian genomes, and thus it is important to extend PCR based assays to cover these regions.

Our approach differs from existing approaches primarily in the evaluation of primer feasibility. Rather than designing an ad hoc primer quality metric, we use a single thermodynamic measure of primer pair quality to identify an acceptable primer pair and a thermodynamically motivated heuristic to ensure that the primers will amplify only the desired locus. In our approach, the user must specify constraints that primers must satisfy (such as a specified range of melting temperatures and lengths), and then we enumerate the acceptable primer pairs that flank a locus, outputting the first acceptable pair according to our primer quality and specificity metrics.

In addition to performance considerations, our approach has several advantages compared to current approaches. First, our assessment of primer pair feasibility is based on thermodynamics; this is in contrast to methods such as Primer3, where primer feasibility is predicted using an ad hoc scoring function. Second, our method requires relatively few free parameters; these parameters are physically meaningful, and thus our method is easier to use.

While the Primer3 primers have a high success rate, our results show that the Primer3 primer assessments are overly conservative, and rejected most of Pythia's primers. The conservative approach does well in most genomic regions but is unable to densely tile challenging regions such as the interspersed repeats in the human genome.

When Primer3 is unable to choose primers for a particular region, users are advised to relax the various quality thresholds [164]. However, it is often unclear how to carry out this relaxation in a principled way. In contrast, our approach uses the minimum equilibrium efficiency, with only one parameter that can be adjusted independently of reaction conditions. We have shown that our approach to assessing primer quality is more accurate than Primer3.

Predicting the specificity of primers is more challenging. The P316 method for assessing primer specificity only examines the number of occurrences of the 16 bases at the 3' end of each primer, and not the genomic arrangement of these occurrences. Taking into account the genomic arrangement of these occurrences is crucial, because two putative primer binding sites must be close and oriented appropriately to generate exponential amplification.

Pythia's specificity predictions are also of limited predictive power: none of the amplicons that Pythia predicted would be non-specific generated multi-modal melting curves. Of course, some of the single peaked melting curves generated multiple products, but it is likely that at least a few of the multi-modal melting curves would have shown multiple bands if their products were run on a gel, and so at least some of the multiple peaked melting curves should have generated multiple peaks as well. The specificity heuristic that we employ will identify obviously bad primers, but it has limitations: Taq polymerases can synthesize strands even with a mismatch at the very 3' base of the primer [196, 133, 94], with strongly sequence dependent efficiency, and can tolerate mismatches near the 3' terminus as well.

Many of the limitations of Pythia stem from an incomplete understanding of DNA stability in PCR mixtures. Many PCR formulations, such as the one used in this study, rely on DNA denaturants that preferentially destabilize GC base pairs. These denaturants improve the success of PCR, especially when amplifying GC rich templates; however, they also significantly distort DNA stability parameters. A better understanding of DNA thermodynamics in the presence of these solvent additives would improve both Pythia's primer acceptability scoring method and Pythia's primer specificity assessments.

Methods

PCRs

PCRs were run using the Immomix master mix, 35 ng human genomic DNA from the GM cell line, and 0.6 μM primers with SYBR green I used as a fluorescent reporter dye. PCRs were run according to the following thermal cycling program: 95C, 7 min, followed by 35 cycles of 98C, 15s; 60 C, 15s; 68C, 45s on an ABI 7900 HT. Each PCR was run twice.

After thermal cycling, a melting curve was taken by slowly increasing the temperature from 68C to 98C and measuring SYBR green I fluorescence. The negative derivative of this fluorescence profile was taken and manually scored according to morphology. All reactions with inconsistent labels among replicates were eliminated from further analysis.

Estimating PCR Success Rates

Under the extrapolation from the stringent success rates, the overall success rate is calculated as
$$S_s = \frac{0.82 * V + 0.67 * I}{V + I}, \tag{4.1}$$
where V is the number of PCRs labeled "valid" and I is the number of PCRs labeled "invalid". Similarly, under extrapolation from the permissive success rates, the overall success rate is calculated as
$$S_p = \frac{0.95 * V + 0.84 * I}{V + I}. \tag{4.2}$$

Acknowldegments

We thank Timothy Rose, Charles Laird, Diane Genereux, and Reinhard Stöger for helpful discussions and comments.

Funding

This work was supported by NIH grant R01 GM071923, and NHGRI grant T32 HG00035.

Legends

Figure 1

Flowchart of the Pythia algorithm. Inputs are the genomic sequence, locus coordinates, and user specified parameters. In step 1, Pythia identifies all primer pairs meeting the user specified requirements and sorts these primer pairs by the sum of the differences between the computed and target primer melting temperatures. In step 2, Pythia computes the thermodynamic quality metric for the top ranked candidate. If this candidate meets a user specified metric threshold, then Pythia proceeds to step 3. If not, the top ranked candidate is removed from the list and Pythia returns to step 2. In step 3, Pythia performs a specificity check. If the primer passes the specificity check, it is given to the user, and the program terminates. If not, the top ranked candidate is removed from the list and Pythia returns to step 2.

Figure 2

Species accounted for in primer feasibility analysis. The solid line is the top strand of the template;the dashed line is the bottom strand of the template; the arrow with the square end is the left primer; the arrow with the round end is the right primer; three dashed lines indicate binding (or folding) via hydrogen bonding. (A) Desired binding interactions. High rates of binding are desired between the primers and the template priming regions. (B) Undesired binding and folding reactions. Primers should not fold, dimerize, or bind to the target outside of the priming regions.

Figure 3

Primer pair design coverages for interspersed repeat regions. Design coverage is defined as the fraction of an interval covered by PCR product sequences. (A) Histogram of coverages for Pythia (mean 80%). (B) Histogram of coverages for P316 (mean 50%).

Table 1:Concordance between gel and melting curves.

For a selected set of PCR primers, we compared the results of melting curve analysis to agarose gel analysis of PCR amplicon. Melting curves were classified as valid or invalid based on melting curve morphology, and gel lanes were classified as clean or not clean at two levels of stringency. In each table entry, the numbers correspond to the number of reactions with the corresponding gel and melting curve label at stringent and permissive levels of gel scoring stringency.

Table 2:Genomic characteristics of selected human genome regions

We compared the ability of Pythia to the ability of the P316 algorithm to tile these regions. We show the location, size, and a brief description of each locus.

Table 3:Primer design performance for selected human regions

Shown are the number of PCRs and the extrapolated success rates for permissive and stringent criteria.

Table 4:Training set sizes

The number of training points for each acceptability threshold. For each threshold, we show the number of examples used to train the SVM, and the ROC and ROC50 scores. We assessed SVM performance using 5-fold cross validation.

Table 5: Pythia acceptability assesments

Pythia acceptability assessment of P316 primers. We assessed the ability of the Pythia primer pair quality metric to predict the quality of the P316 primers. Here, we compared the Pythia primer assessment to the results of melting curve analysis.

Table 6:P316 acceptability assesments

P316 acceptability assessment of Pythia primers.We assessed the ability of the Primer3 scoring function to predict the quality of the Pythia primers. Here we compare the P316 primer assessment to the results of melting curve analysis.

Table 7: Pythia specificity assessments

Pythia specificity assessment of P316 primers. We assessed the ability of the Pythia specificity assessment to predict which amplicons would yield non-specific products. Shown is the results of the Pythia specificity assessment of P316 primers compared to the labels of melting curves.

Table 8: P316 specificity assesments

P316 specificity assessment of Pythia primers. We assessed the ability of the P316 specificity assessment to predict which amplicons would yield non-specific products. Shown is the results of the P316 specificity assessment of Pythia primers compared to the labels of melting curves.

Figures

Tables

Chapter 5
MULTIPLE TARGET PRIMER DESIGN

5.1 Introduction

The systematic study of large collections of genomic elements is a key goal of functional genomics. However, reagents can be prohibitively expensive for large scale studies, especially when each genomic element must be assayed individually. This problem is especially acute when the polymerase chain reaction (PCR) is used to interrogate genomic elements, because two DNA sequences (the primer pair) must be specifically designed and synthesized for each reaction. In order to reduce these reagent costs, we developed a strategy for designing primer pairs that each amplify more than one genomic element.

Our strategy requires that several distinct amplicons can be tolerated in each reaction. In this work, we focus on designing primers for ascertaining cytosine methylation patterns in human genomic DNA. In these epigenetic studies, PCR products are cloned into bacteria and then plasmid inserts are sequenced from individual colonies [25, 187]. Because we sequence clones from PCR products, we reduce primer costs by using each PCR to amplify several distinct templates.

Typically, when a primer pair must amplify distinct templates, the templates are homologous and well conserved [102, 160]. Because these multi-target primers are conserved, good multiple alignments can be used to design primers that are highly similar in sequence to each of the targets [102, 160, 84, 57]. These multi-target primers are typically designed to be similar to the consensus sequence generated by the multiple alignment, and often include degenerate positions at ambiguous positions in the multiple alignment to ensure that a sub-population of the primer molecules will have an exact match to each of the targets.

In the absence of conservation between sequences, however, the multiple alignment approach is not applicable. We therefore developed a strategy to design primers to amplify multiple genomic elements in the absence of sequence conservation, and hence the absence of good multiple alignments. In our approach, we use thermodynamic models of DNA binding rather than a multiple alignment to design primers that bind stably to their targets.

Stable binding to the target is not the only requirement for successful priming; a primer must have high similarity at the 3' end of the primer sequence. The primers must have high binding affinity to the targets so that a substantial fraction of targets will be bound by the primers (a prerequisite for successful PCR), and they must have high 3' similarity because the efficiency of primer extension by a polymerase can be sensitive to the degree of complementarity at the 3' end of the primer [94, 133]. Our approach relies upon the hypothesis that an overall level of affinity, coupled with nearly exact complementarity at a short region of the primer 3' end, is sufficient for priming success.

Thus, given a collection of genomic elements, we seek to design a small set of primers that collectively amplifies all of the target elements. This problem is similar to problems known to be NP-complete [102], and thus we use heuristics to achieve acceptable solutions. The heuristic that we employ reduces the difficulty of designing a DNA primer sequence to bind to a set of target sequences. Our heuristic works by identifying sets of thermodynamically homogeneous sequences; by choosing thermodynamically similar sequences, we significantly reduce the difficulty of the primer design problem.

We developed a randomized approach to multi-target primer design. The input to our approach are a set of loci, and the output is a set of primers that collectively amplify all targets in the input set. Our algorithm is summarized in Figure 5.1. First, we randomly choose a locus in the input set that is not yet targeted by primer pairs.

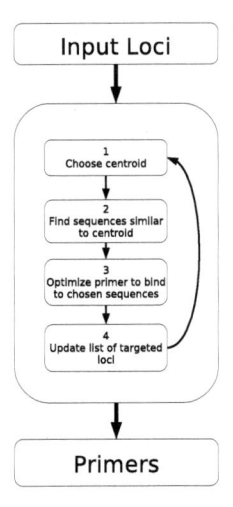

Figure 5.1: Algorithm for multiple target primer design. The input are a set of loci, and the output is a set of primers that will prime each of the loci; our objective is to make the primer set as small as possible. The algorithm works in 4 steps. 1: A sequence not targeted by a primer, call the centroid, is chosen. 2: Untargeted loci similar to the centroid are identified. 3: Primers are optimized for binding affinity to each element of the set. 4: The list of targeted elements is updated based on the sequences to which the optimized primers will stably bind.

We call this locus the centroid. We identify two candidate priming targets at the 5' and 3' ends of that locus. In step 2, we examine all of the other loci and identify those elements that have candidate priming targets that are thermodynamically similar to the centroid primer targets. In step 3, we optimize a sequence that will bind stably to the candidates, using the centroid primer target sequences as starting points in the optimization. Finally, in step 4, we update our list of targeted loci according to which elements the optimized primers could bind stably. We repeat steps 1-4 for a specified number of epochs, and then design standard single target primers for any loci that are not yet targeted by multiple target primers.

Our approach relies on our ability to optimize a sequence to bind to each member of a set of sequences, in step 3 of Figure 5.1. However, this optimization problem is difficult because it may not always be possible to find a sequence that will bind well to each member of a sequence set. In order to address this difficulty, we describe a method to identify thermodynamically homogeneous pairs of sequence sets; these sets have the property that the sequences in each set have similar binding partners. Because the sequences all have similar binding partners to one another, we greatly increase our chances of finding a primer sequence that can bind to multiple elements in the set.

We focus in particular on designing a small set of primer pairs that will each amplify loci from CpG islands for epigenetic analysis. Given approximately 478 loci of interest, we designed primers to facilitate methylation analysis of CpG dinucleotides. We designed 205 primers that target all 478 loci, resulting in substantial savings in reagent costs.

Further, we show that our heuristic is very effective at identifying loci in which both the left and the right primer candidates are thermodynamically homogeneous, and that this homogeneity substantially improves the success rate of the optimization of the primer sequences for binding to the target sequences.

5.2 Methylation Assay

In this work, we design primers for amplification of bisulfite converted DNA for CpG methylation analysis. CpG methylation is an epigenetic phenomena associated with silencing of gene expression [154]. CpG methylation can be detected by using sodium bisulfite conversion. Sodium bisulfite converts unmethylated cytosine bases to uracil bases, but leaves methylated cytosines unchanged [207]. By subjecting a DNA sample to bisulfite conversion and then sequencing cloned PCR products, methlyated cytosines can be identified in individual molecules [34].

The bisulfite conversion assay is depicted in Figure 5.2. First, genomic DNA is cut with a restriction enzyme that cuts once in a locus of interest. For this assay, the restriction enzymes should be insensitive to CpG methylation and leave a staggered cut-site so that a linker sequence can be ligated to the cut DNA. Next, a hairpin linker is ligated to the cut DNA. This hairpin linker is used to ensure that DNA samples are uncontaminated and that sample molecules are not repeatedly analyzed after cloning [123]; in addition, the hairpin linker assay allows hemi-methylation (i.e. methylation on only one strand) to be detected [25]. DNA is then subjected to bisulfite conversion, PCR amplified, cloned, and sequenced. Comparison of experimentally obtained sequences to the reference sequence then reveals which cytosines were methylated.

Because the sequences both 5' and 3' of a restriction enzyme cut-site typically contain CpG dinucleotides, primers must be designed to amplify these two sub-loci for each locus with a restriction enzyme recognition site. Thus, given n loci of interest, $2n$ primer pairs must be designed for bisulfite conversion.

Sodium bisulfite conversion has two important impacts on primer design for bisulfite converted sequences. First, because sodium bisulfite conversion changes unmethlyated cytosines to uracils, bisulfite converted DNA has a strong sequence bias away from cytosine nucleotides and toward adenine and thymine bases. Thus, primers are easier to design due to a strongly biased sequence composition. Second, AT base

83

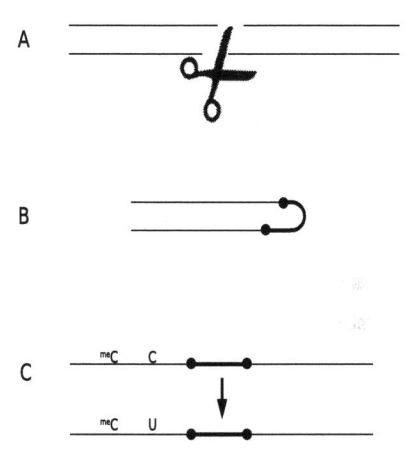

Figure 5.2: Bisulfite conversion assay. (A) First, genomic DNA is cut with a methylation insensitive restriction enzyme that cuts in a locus of interest. (B) A hairpin linker is then ligated to the cut-site. (C) the DNA is then subjected to bisulfite conversion, which changes unmethylated cytosines to uracil, but leaves methylated cytosines unchanged. These samples are then PCR amplified, cloned, and sequenced to identify methlyated cytosines.

pairs are less stable than GC base pairs [83]; thus it is harder to design an imperfectly complementary sequence to bind stably to an AT rich sequence than a GC rich sequence.

5.3 Algorithms

Our approach is illustrated in Figure 5.3. First, we enumerate all ordered pairs of 5-mers, and identify all loci in which the first word occurs 5' of the second word in a sequence subjected to in-silico digestion, ligation, and bisulfite conversion. The sequences that a primer must target for each locus and word pair are the reverse complement 30 base pair sequence ending at the first word, and the 30 base pair sequence beginning at the second word. Loci are discarded if either candidate primer target contains a CpG dinucleotide.

Given this dictionary of loci, in which each locus appears multiple times, we use a randomized algorithm to design primers to target subsets of the loci as follows. In step one, a locus is chosen at random and designated as the centroid; this is the first element in the set of candidate targets. Next, all other loci are compared to the centroid, and added to the set if there is at most one mismatch between the 5 bases at the 3' end of target terminus and the centroid, and if the thermodynamic similarity between to the centroid is above a threshold. We describe our thermodynamic similarity metric below.

In step two, a pair of graphs is formed whose nodes are the left and right targets. Two nodes in a graph are connected by an edge if their thermodynamic similarity is above a threshold. In this step, loci are identified with the property that their left sequences form a clique in the left target graph, and their right sequences form a clique in the right target graph. These cliques are identified using a one-class support vector machine [177].

In step three, the left primer sequence is optimized to bind to the reverse complements of the sequences in the left clique and the right primer sequence is optimized

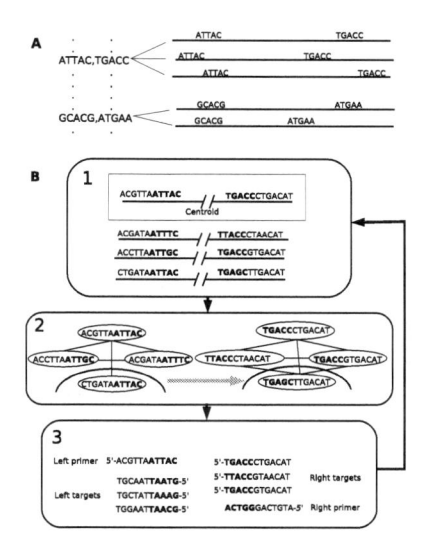

Figure 5.3: Illustration of our algorithm for multiple target primer design. The input is a set of sequences for which primers must be designed, and the output is a set of primers that amplify multiple elements in the set. (A) We enumerate all ordered pairs of 5-mers and identify all loci in which the first word occurs 5' of the second word. The sequences that a primer must target for each locus and word pair are the reverse complement of the 30 base pair sequence ending at the first word, and the 30 base pair sequence beginning at the second word. Loci are discarded if either candidate primer target contains a CpG dinucleotide. (B) Our primer design algorithm has three parts. First, a locus is chosen at random and designated as the centroid; this is the first element in the set of candidate targets. All other loci are compared to the centroid, and added to the set if there is at most one mismatch between each target terminus and the centroid, and if the thermodynamic similarity between a locus and the centroid, according to our thermodynamic similarity function, is above a threshold. Second, two graphs are formed whose nodes are the left and right targets. Two nodes in a graph are connected by an edge if their thermodynamic similarity is above a threshold. In this step, loci are identified with the property that their left sequences form a clique in the left target graph, and their right sequences form a clique in the right target graph. These cliques are identified using a one-class support vector machine. Finally, primer sequences are optimized to bind to the reverse complements of the sequences in the left clique and to the sequences in the right clique.

to bind to the sequences in the right clique. We use simulated annealing for this optimization, and we use the left centroid as the starting sequence for optimization for the left sequences, and the reverse complement of the right centroid as the starting sequence for optimization for the right sequences. We then identify those sequences whose left sequences and whose right sequences have at least a specified level of binding affinity to the resulting primer sequences as the loci targeted by a primer pair.

Thermodynamic similarity is crucially important for this work. We therefore present our concept of thermodynamic similarity in the next section, and then describe how to identify groups of thermodynamically similar sequences using a one-class support vector machine. Finally, we describe how we use simulated annealing to optimize binding efficiency.

5.3.1 Thermodynamic homogeneity

We define the thermodynamic similarity of two sequences as the fraction of one bound sequence to the reverse complement of the other (or vice-versa), at thermodynamic equilibrium in a system where both strands were present at equal initial concentrations. In this model system, the only reaction considered is binding between one sequence and the reverse complement of the other; DNA folding, and self-dimerization are neglected.

In order to compute this fraction, we first compute the free energy of binding for one sequence binding to the reverse complement of the other and then compute the fraction of molecules in the dimer form at thermodynamic equilibrium. Figure 5.3.1 illustrates this definition. When two sequences are similar under this definition, then they will bind to similar sets of sequences, and thus it is easier to find a sequence that will bind to each.

We use statistical mechanical models of DNA binding [45, 60] to compute the free energy of binding between two sequences. These models evaluate the stability of a wide variety of conformations in which one molecule is bound to the other via

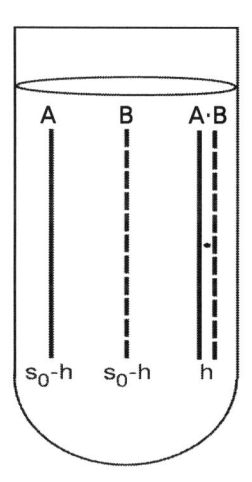

Figure 5.4: Our thermodynamic definition of similarity is based on the fraction of strands of sequence A (solid line) bound to the reverse complement of sequence B (dashed line) for a system in which the strands were present at equal initial concentrations s_0. If A is thermodynamically similar to B, then at equilibrium the concentration h of dimers will be high, and the concentrations of the single strands $s_0 - h$ will be low.

hydrogen bonds, and then sum the stabilities of all of these conformations using a dynamic programming algorithm. These models rely on thermodynamic reference data which specifies the stability of features such as internal loops, base pairing, and stacking; we use the parameter set described in Santa Lucia et al. [83].

Our thermodynamic similarity function relies on basic analysis of the chemistry of binding reactions. Given a free energy of binding between two DNA sequences a and b, the fraction of sequences in the dimer form at equilibrium is,

$$f(a,b) = \frac{2s_0 + \frac{1}{K} - \sqrt{(2s_0 + \frac{1}{K})^2 - 4s_0^2}}{2s_0}, \qquad (5.1)$$

where $K = e^{-\Delta G/RT}$, and both strands are present in equal concentrations (s_0 moles per liter). We use this measure of the fraction of bound strands to define a similarity function s on DNA sequences s_i and s_j as

$$s(s_i, s_j) = f(s_i, R(s_j)) + f(s_j, R(s_i)) \qquad (5.2)$$

where R is the reverse complement function.

A set of sequences is thermodynamically homogeneous when they have similar binding partners; this relationship can be quantified by the degree of equilibrium binding between each sequence and the reverse complements of all the others. The intuition underlying this work is that when sequences have similar binding partners, then it is easier to design a sequence to bind to each of them than when their binding partners are dissimilar.

5.3.2 Identification of thermodynamically homogeneous sequences

Given a set of sequences, we seek a subset that are all thermodynamically similar to one another. To do this, we compute the similarity of each sequence to all the others, and we collect these values into a matrix $\hat{K}$, defined as

$$\hat{K}(i,j) = \text{similarity}(s_i, s_j). \tag{5.3}$$

In order to employ our scheme for identification of thermodynamically homogeneous sequences, we then form a matrix K by adding a positive constant to each element of the diagonal in order to make the matrix positive semidefinite. Because the resulting matrix is positive semidefinite, there exists a function $\Phi(s_i)$ that maps each sequence s_i to a point in a Hilbert space [40], and each element of the similarity matrix $K(i,j)$ is the inner product of the images of sequence s_i and s_j under the mapping Φ, i.e.

$$K(i,j) = \langle \Phi(s_i), \Phi(s_j) \rangle \tag{5.4}$$

We use this geometric interpretation of the elements of the similarity matrix K to identify thermodynamically homogeneous sequences by using a one class SVM. Given a set of points in a Hilbert space, as illustrated in Figure 5.3.2, the one class SVM [177] attempts to find a small region of the space that contains most of the input points. The one class SVM takes as input a kernel matrix, whose entries are the inner products of the points in the feature space, and solves the following quadratic program:

$$\min_{\alpha \in R^M} \frac{1}{2} \sum_{i,j} \alpha_i \alpha_j K(i,j), \text{subject to } 0 \leq \alpha_i \leq \frac{1}{\nu M}, \sum \alpha_i = 1, \tag{5.5}$$

where M is the number of training examples, and ν is the pre-specified fraction of outliers. The solution of this quadratic program determines a hyperplane that separates a small group of points from the origin; these points are the thermodynamically homogeneous sequences. Because the similarity measure gives a number strictly between 0 and 1, and we use it in conditions in which most strands will be bound, the data can be regarded as located close to the surface of a hypersphere. Separating the

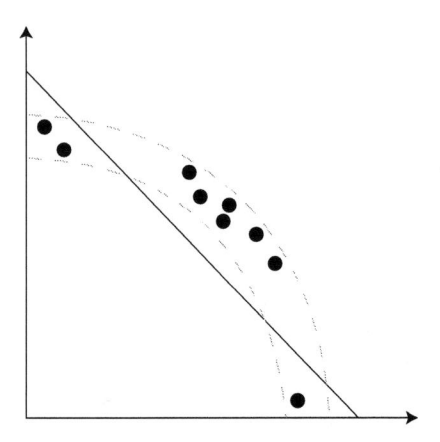

Figure 5.5: Our similarity function produces a similarity metric that lies strictly between 0 and 1. Because we use it at temperatures such that most strands will be in the dimer form for a sequence and its reverse complement, the norms of the data under the embedding are all close to 1, and the data can be regarded as located near the surface of a hypersphere. In this context, training the SVM to separate data from the origin is equivalent to finding a small ball that contains most of the data.

data from the origin then corresponds to finding a small ball that contains most of the data in the Hilbert space.

Due to the construction of the kernel matrix, proximity in the feature space corresponds to similarity of binding partners among sequences. Therefore, the sequences defined by this procedure comprise a set of thermodynamically homogeneous sequences. We show below that it is easier to optimize a sequence to bind to each of these inliers than to the whole input set.

We can further improve our chances of designing good primer pairs by finding sequence sets which are jointly homogeneous in the left and the right sequences. Let K_l be the kernel matrix for the left set, and K_r be the kernel matrix for the right set of sequences. Then, the matrix

$$K_b = K_l + K_r \tag{5.6}$$

is a kernel matrix (due to closure of kernel matrices under summation [40]) whose entries will be large when both the left pair and the right pair of sequences have similar binding partners. Training a one-class SVM on this joint kernel matrix identifies a set of sequences that are thermodynamically homogeneous in the left and right sequences, and we show below that by finding jointly similar sets of sequences, we significantly increase the number of primable sequences.

One important issue in the use of a one-class SVM is the choice of the fraction of outliers. The kernel framework we employ facilitates a geometric analysis of the problem of optimizing a sequence to bind well to each member of a set. In the feature space, the thermodynamic similarity of two distinct points will be inversely proportional to the distance between them, because they all have similar norms. In order to choose the fraction of outliers, we pre-specify a minimum level of mutual similarity, and then raise the outlier fraction until the minimum thermodynamic similarity of any two elements in the inlier set exceeds the threshold.

5.3.3 Simulated Annealing Optimization

Given a set of sequences with similar subsequences at the 5' and 3' end, we must design primers to bind stably to the primer targets. We use simulated annealing [91] to find sequences that will bind well to the input set. Our objective function $O(c, S)$ is simply the mean free energy of binding, ΔG, between a candidate sequence c and each sequence s_i of the input set S:

$$O(c, S) = \frac{1}{|S|} \sum_i \Delta G(s_i, c), \tag{5.7}$$

We use simulated annealing to choose a sequence that minimizes this objective function.

Simulated annealing is a simple optimization procedure that uses a local search to iteratively improve a candidate solution. Given such a candidate solution, simulated annealing perturbs it, accepting the updated sequences if they decrease the objective function, and accepting the updated sequences with a probability inversely proportional to the amount of increase if they do not improve the objective function. In our simulated annealing procedure, moves are accepted with probability

$$p = \min(1, e^{-\frac{O(c_i, S) - O(c_{i-1}, S)}{T_i}}) \tag{5.8}$$

where c_i is the sequence at epoch i of the optimization, and T_i is the temperature at epoch i. The temperature, which decreases according to the exponential schedule

$$T_i = cT_{i-1}, \tag{5.9}$$

determines how large a decrease in the objective function is likely to be accepted. In the beginning of the procedure, at high temperature, many perturbations will be accepted, whereas at the end of the procedure, only those that improve the objective

Table 5.1: Methylation Insensitive Restriction Enzymes.

Enzyme	Recognition Site
BsawI	WCCGGW
SphI	GCATGC
SacI	GAGCTC
KpnI	GGTACC
BssSI	CACGAG
SpeI	ACTAGT
AflII	GTTAAG
AvrII	CCTAGG
BclI	TGATCA
PstI	CTGCAG
BglII	AGATCT

function will be accepted.

Our annealing procedure uses three kinds of perturbations. Given a sequence to perturb, with probability 1/2 we change a base in the sequence to a new base; with probability 1/4 we insert a random base at a random position and remove a base from the beginning or end with equal probability; and with probability 1/4 we delete a base at random and add a random base to the beginning or end with equal probability.

5.4 Methods

We focused on a set of 500 loci consisting of CpG islands overlapping promoter elements in the human genome. We simulated restriction digestion of these loci for a set of methylation insensitive restriction enzymes, linker ligation, and then bisulfite conversion. We then used the resulting set of sequences to evaluate our multiple target primer design method.

5.4.1 Methylation Insensitive Restriction Enzymes

Due to the experimental protocol we use, we require each locus to have a restriction enzyme cut-size. We therefore chose a set of restriction enzymes such that each enzyme was insensitive to cytosine methylation and no enzymes yielded blunt ended cut sites. These enzymes are presented in Table 5.1. We examined our 500 loci for cut-sites, and found that 487 loci had at least one cut site. We then simulated digestion, ligation, and bisulfite conversion, and used the resulting sequences for evaluation of our multiple target primer design.

5.4.2 Evaluating Similarity

We evaluated the effects of varying levels of centroid similarity and mutual similarity on our algorithm. Using the set of loci we obtained by simulating digestion, ligation, and conversion, we used our algorithm to design multiple target primers while varying the level of locus similarity to the centroid, and the level of mutual locus similarity of the resulting set. For each level of centroid similarity, we attempted to find a subset with at least four elements with a specified level of mutual similarity using the SVM algorithm. If we were able to find such a subset, we then chose a random subset of the same size, in order to verify that the difficulty of multiple target primer design is not simply a function of the size of the inputs. For each set, we compared the number of primable sequences.

5.4.3 Evaluating the Multiple Target Primer Design Algorithm

We also evaluated the overall performance of our algorithm on the CpG island loci. We enumerated all pairs of 5-mers and collected all sequence sets with at least 2 members, with one 5-mer appearing between 100 and 350 bases 5' of the other 5-mer in each set. We then iterated our algorithm and evaluated how many primer pairs we needed to generate at least one product in each of the 478 loci with at least one

Table 5.2: Mean number of primable sequences for varying joint and centroid similarity. Each image is a heat map, with the joint similarity on the x-axis and the centroid similarity on the y-axis.

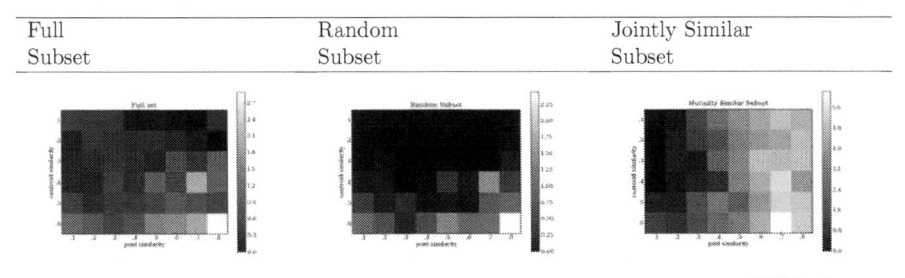

Full Subset	Random Subset	Jointly Similar Subset

restriction cut site. We further evaluated the number of primers required to generate products at all loci both 5' and 3' of any cut site. We then evaluated the fraction of CpG dinucleotides that were within 175 bases of a cutsite, and the fraction of these that could be assayed using primers developed using this approach.

5.5 Results

We evaluated our multiple target primer design algorithm on a 500 locus set of CpG islands overlapping gene promoters. We show that mutual similarity is more important than centroid similarity when designing multiple target primers, and we show that we can reduce the cost of primers by 50% as compared to designing one primer pair per locus.

5.5.1 Centroid Similarity and Joint Similarity

Table 5.2 shows that ensuring joint similarity among the candidate targets is more important than ensuring similarity to the centroid. When optimizing primers to bind to the full set and the random subset, the best performance was 2.5 primable sequences on average, and that was only when a highly similar subset could be identified by the

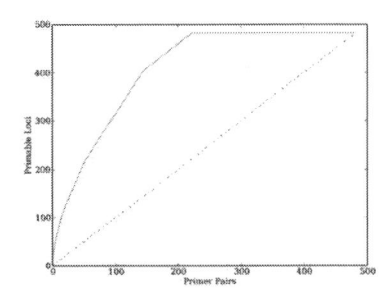

Figure 5.6: Number of primable loci as a function of primer pairs. Dashed line is the coverage assuming one primer pair per locus ($y = x$); Solid line is number of primable loci using multi-target primers. All loci can be primed using 204 primer pairs.

SVM in the full set. In contrast, the mean number of primable sequences was not strongly sensitive to the centroid similarity but was sensitive to the mutual similarity. On average, about 5 sequences were primable for mutual similarities above 0.6, over the full range of centroid similarities evaluated.

5.5.2 Coverage of the 500 CpG Islands

We were able to place at least one amplicon in all loci using 204 primers pairs. Figure 5.5.2 shows the number of primable loci as a function of primer pairs. This list of 204 primer pairs was distilled from a set of 11,000 candidates iteratively as follows. First, the primer pair with the most amplicons in unprimed loci was selected, and that locus was removed from the set of primable loci for all other primer pairs. The list was re-sorted, and the pair with the most remaining unprimed loci selected, until all primers targeting multiple loci were removed.

There were approximately 1250 cut sites in the 500 loci, generating approximately 2500 sequences to prime. Figure 5.5.2 shows the number of primable loci as a function

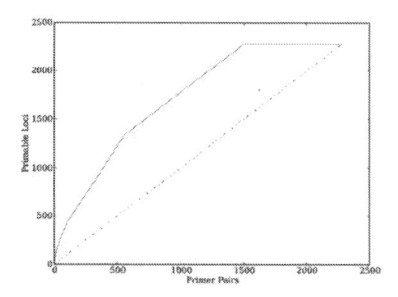

Figure 5.7: Number of primable cut sites (with the 5' segment treated as distinct from the 3' segment) as a function of primer pairs. Dashed line is the coverage assuming one primer pair per locus ($y = x$); Solid line is number of primable loci using multi-target primers. All 2500 loci can be primed using approximately 1500 primer pairs.

of the number primer pairs. We are able to design primers to cover this set using 1500 primers. However, using a set of 500 primers, we can target approximately 1500 loci. We generated this set using the same algorithm used to generate the set summarized by Figure 5.5.2 and the same set of primer candidates.

Both curves in Figure 5.5.2 and Figure 5.5.2 show distinct line segments in the shoulders. These line segments are artifacts of the way that the curves were generated. The line segments appear in the curve because coverage was plotted in order of the size of the set of targets that each primer covered. Thus, all primers covering two unique loci are plotted together, and are responsible for the line segment with the slope of 2, and the set of loci not covered by any multiple target primer (and thus for which single target primers must be designed) are also plotted together, generating the line segment with the slope of 1.

5.6 Discussion

We addressed the problem of designing a small set of primers to prime a large set of unrelated genomic elements. Our approach can result in at least a two-fold cost reduction in primer design costs. These savings can likely be improved with further application of our randomized algorithm. As the scale of functional genomic studies increases, these savings in reagent costs will become increasingly significant.

We have shown that, given a set of sequences for which primer pairs must be designed, extracting the sequences that are thermodynamically homogeneous and optimizing primers to bind to them is substantially more effective than optimizing over the full input set. Our one-class SVM is able to choose subsets for which the required sequence optimization is effective on the basis of a geometric analysis of DNA binding interactions for a particular set of sequences.

Designing primers to amplify bisulfite converted CpG islands is a challenging primer design problem, because the bisulfite conversion process results in a significant skew of nucleotide composition toward AT base pairs. This skew is exacerbated by the requirement that the primers should not overlap CpG dinucleotides (which are the candidates for methylation).

This approach awaits experimental validation. Such experiments were in progress but not completed in time to be included in this thesis. One complication is the possibility of mismatches within five bases of the 3' terminus. Results from the polymerase literature suggest that polymerase enzymes can extend a primer even in the presence of a limited number of mismatches at the 3' end. However, if that proves to be problematic, then degenerate primers can be designed with ambiguous nucleotides at the positions at the 3' primer terminus so that every primer target has an exact 3' match for a subset of the primer pool.

Chapter 6

A TEST OF MULTIPLE TARGET PRIMER DESIGN

6.1 Introduction

In order to test our fundamental hypothesis that significant overall binding affinity and a short stretch of exact 3' agreement between the primers and their targets are sufficient for PCR by Taq polymerase, we used the *E. coli* bacterial genome as a model system for our multi-target primer design. We designed primers that should each amplify multiple targets in *E. coli*, performed PCRs and compared the results to our predictions.

We used a longer region of 3' agreement for this study than in chapter 5; all of the primers tested here had 9 exactly agreeing bases at the 3' end. We changed the length of 3' agreement due to preliminary experiments with short primers that did not reliably amplify products. The primers in the preliminary experiments were much shorter than the primers used in the CpG design in chapter 5 (16 as compared to 30 bases). Because these short primers were ineffective, we increased both the overall length and the length of the exactly agreeing 3' primer ends.

We designed three primer pairs that should each amplify three targets in *E. coli*. The PCR products for each primer pair were separated in length by at least 90 bases, so that they could all be clearly resolvable on an agarose gel when all three products were generated in one reaction. We tested our approach both by assessing the ability of the multiple target primers to amplify purified PCR products, and also by purifying gel bands, cloning PCR products, and then sequencing clones.

We were able to achieve amplification for at least 6 of the 9 targets from purified PCR products. In 2 of the 3 cases, the least stable primer pair was the one with no visible product. The cloning was less successful, although we were able to sequence at least one clone from each primer pair.

6.2 Primer Design

We designed our primers by first enumerating all pairs of 9-mers in the *E. coli* genome such that each pair of 9-mers had at least three occurrences where the 9-mers were separated by at most 600 base pairs.

We extracted the 25 base sequence ending at the left 9-mer, and the 25 base sequence beginning at the right 9-mer. We then attempted to optimize primer sequences to bind to the right sequences and the reverse complements of the left sequences using the simulated annealing approach described in Chapter 5.

We chose three sets of primer pairs whose products were separated in length, and whose free energies of binding between the primers and their targets were such that the shortest amplicon would have the most stable overall pair of binding affinities, the middle amplicon would have an intermediate level of binding affinities, and the longest amplicon would have the least binding affinity between the primer binding sites and the primers.

Table 6.1 shows the products, their lengths, and the free energies for the binding sites at 55 C and 45 C. We found that the reactions worked best at 45 C, where the free energies were substantially greater than at 55 C. Table 6.2 shows the primer sequences used, and Table 6.3 shows the genomic coordinates of the expected PCR products for each primer pair.

6.3 Methods

We performed PCRs for the cloning experiments as follows. We used 1.25 units of HotStartTaq Polymerase (Qiagen) in 50 microliter volumes with 50 mM NaCl$_2$, 2 mM MgCl$_2$, 0.8 mM dNTP, 0.1 micromolar primer concentration. We used $2 \cdot 10^4$ initial genomic copies. PCR thermal cycles were as follows: 95 C for 5 min; 95C for 1 min

Table 6.1: Expected products. Column 1 is the primer pair ID. Column 2 is the product ID (3 for each primer pair). Columns 3 and 4 are the free energies of binding between the left primer and the target at 55 C and 45 C, respectively. Columns 5 and 6 are the free energies of binding between the right primer and its target at 55 C and 45 C, respectively. Column 7 is the product length.

Primer Pair	Locus ID	Left ΔG (55 C)	Left ΔG (45 C)	Right ΔG (55 C)	Right ΔG (45 C)	Product Length
M1	E1	-11.0	-14.6	-12.9	-17.1	233
M1	E2	-11.4	-14.4	-11.5	-14.5	321
M1	E3	-9.1	-11.5	-9.1	-12.5	492
M2	E4	-11.0	-15.1	-15.0	-18.6	192
M2	E5	-12.8	-16.8	-10.9	-15.9	297
M2	E6	-11.0	-14.5	-8.75	-12.9	474
M3	E7	-14.9	-19.2	-11.3	-15.1	352
M3	E8	-8.3	-10.9	-12.6	-16.9	453
M3	E9	-11.1	-14.6	-8.3	-10.8	577

Table 6.2: Primer sequences used for experiments.

Primer Pair	Left Primer	Right Primer
1	GGGCGCATAGGAGGCACGCAGCAGA	CATGCGTGGTGTGCTCCGCTGCTGT
2	TGCGCGCTGGGCGAAGCGCTGCTGA	TTTGAGCGCGCGGCTGATGATCAGC
3	GTGACTAAGCTGGCGGCTGACGCTG	ATTCACGGACGCACCTGCAGCATCG

Table 6.3: Genomic coordinates of the expected products.

Primer Pair	Locus ID	Beginning Base Coordinate	Ending Base Coordinate
M1	E1	3374738	3374946
M1	E2	2349574	2349870
M1	E3	2117156	2117623
M2	E4	3222895	3223062
M2	E5	3673513	3673785
M2	E6	3248802	3249251
M3	E7	130029	130356
M3	E8	1708979	1709407
M3	E9	3605609	3606161

Table 6.4: Primer sequences for resequencing analysis.

Primer ID	Left Primer	Right Primer	Locus contained in amplicon
RE1	GTTAAAGCGCACCTCATCAT	CATCATCAACGGTTCAGCTT	E1
RE2	TGGTCACCAGTGTCATAAAGA	CGGTGATGTTTGTACTGTTGT	E2
RE3	AAGTAACTCAGAATGGTGTTGATAATTT	CACCGTTGATTAAAACATTATCGC	E3
RE4	ACCCTATCGCCACTTATCAG	CTGATCGTATTCGCCGTTAATT	E4
RE5	ACCGATTGGCTTCTTCTTTG	AGACCCATCAGACTTAACCC	E5
RE6	ATGCCGATTCTGTTTATTACCG	ATTCATTTTGATAGAGCGTGTCTC	E6
RE7	GGACTGATAACGCTGACGAG	CCTGATGTGGTCTATATTCCCA	E7
RE8	GATTGTACGCTCAGACGTTAC	GTTTTGCTTTATTCATCAGACATTCC	E8
RE9	GTTGGAAGTGCTGTCAGAAC	TTAATACCGTCACCGACCAT	E9

followed by 60 min annealing at 45 C, for two cycles; 35 cycles of 95 C, 1 min; 60 C, 1 min; 72 C, 1 min; finally 72 C, 1 min. The very long initial annealing/extension steps were found to significantly improve product yield.

Bands were extracted from gels and cloned using the TOPO-TA vectors (Invitrogen) into competent cells. Cells were incubated 16 hours, and then colonies were picked for sequencing. Vectors were amplified from bacterial cells using the HotStarTaq master mix (Qiagen) in 15 microliter volume, and then PCR products were sequenced on an ABI 7700 sequencer using the BigDye2 sequencing reagents (ABI).

Sequences were then searched against the bacterial genomes using BLAST [8], and checked for location and sequence identity to the expected products.

In addition, we designed primers whose amplicons subsumed the expected multiple target amplicons; these primers are shown in Table 6.4. The amplicons for these primers extended at least 50 bases from each end of the multiple target amplicons. These primers were designed with the Pythia software.

We used PCR to amplify these amplicons as follows. We used HotStarTaq master mix (Qiagen) in 50 microliter volumes, with 0.5 micromolar primer concentrations $2 \cdot 10^4$ initial genomic copies, using the following PCR schedule: 95 C for 15 min; 35 cycles of 95 C for 45 sec, 57 C for 45 sec, 72 C for 45 sec; and finally 72 C for 5 minutes.

We then used Microclean (Microzone) to purify the PCR products, and sequenced them as described above. We diluted purified PCR products 100,000 fold, and used 1

Table 6.5: PCR products recovered.

Primer Pair	Product 1	Product 2	Product 3
E1	N	Y	Y
E2	N	Y	N
E3	Y	N	N

microliter of the diluted product as template when amplifying purified PCR products using the multiple target primers.

When amplifying from purified products, we used HotStarTaq master mix in 50 microliter volumes, with primers at 0.5 microliter concentration and 1 microliter of diluted PCR product. We used the following PCR temperature schedule: 95 C for 5 min; 95C for 1 min followed by 60 min annealing at 45 C, for two cycles; 35 cycles of 95 C, 1 min; 57 C, 1 min; 72 C, 1 min; finally 72 C, 1 min.

6.4 Results

We were able to sequence at least one clone for each PCR product. Table 6.5 shows the products recovered for each primer pair in the cloning experiments. In *E. coli* genomic DNA, primer pairs M1 and M3 yielded extra, unexpected bands. Primer pair M2 yielded only two bands. Most of the recovered sequences mapped to regions of the *E. coli* genome with no significantly stable (better than -10 kcal/mole) primer binding sites according to the the thermodynamic models of DNA binding used in this thesis.

We then sought to evaluate whether the primers could amplify products from purified loci. We used the primers in Table 6.4 to generate amplicons from *E. coli* genomic DNA. These products were sequenced, revealing that the targets for the multiple target primers agreed exactly with the *E. coli* reference sequence. We then used purified, diluted PCR products for each locus as starting templates for the

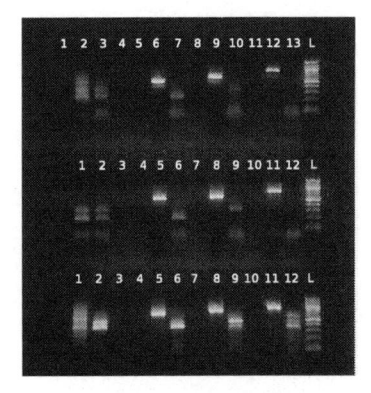

Figure 6.1: Gel showing results of amplification from purified PCR products. See text for explanation of gel lanes. Lanes marked "L" are 100 base pair ladders.

multiple target primers.

The multi-target primers were able to amplify products from purified PCR products that contained their targets. Figure 6.4 shows the results of PCR on purified products and on genomic DNA. Table 6.6 identifies the template and primers used for each lane. Figure 6.4 shows that the multiple target primers are able to generate template from PCR products, except for those with the least stable binding sites (lanes 11-13, row 1; 10-12 rows 2), with the possible exception of the primer pair M3. The templates for these reactions were contaminated with one another (i.e. purified product E7 was shown with PCR to contain templates E8 and E9; likewise both E8 and E9 contained the other two). Thus, M3 clearly amplified something from these products, but it is unclear exactly which products were amplified from this experiment.

Table 6.6: Primers and templates for gel lanes.

Row	Lane	Template	Primers
1	1	*E. Coli genomic DNA*	None
1	2	*E. Coli genomic DNA*	M1
1	3	Purified loci E1-3	M1
1	4	Purified loci E1-3	None
1	5	Purified locus E1	None
1	6	Purified locus E1	RE1
1	7	Purified locus E1	M1
1	8	Purified locus E2	None
1	9	Purified locus E2	RE2
1	10	Purified locus E2	M1
1	11	Purified locus E3	None
1	12	Purified locus E3	RE3
1	13	Purified locus E3	M1
2	1	*E. Coli genomic DNA*	M2
2	2	Purified loci E4-6	M2
2	3	Purified loci E4-6	None
2	4	Purified locus E4	None
2	5	Purified locus E4	RE4
2	6	Purified locus E4	M2
2	7	Purified locus E5	None
2	8	Purified locus E5	RE5
2	9	Purified locus E5	M2
2	10	Purified locus E6	None
2	11	Purified locus E6	RE6
2	12	Purified locus E6	M2
3	1	*E. Coli genomic DNA*	M3
3	2	Purified loci E7-9	M3
3	3	Purified loci E7-9	None
3	4	Purified locus E7	None
3	5	Purified locus E7	RE7
3	6	Purified locus E7	M3
3	7	Purified locus E8	None
3	8	Purified locus E8	RE8
3	9	Purified locus E8	M3
3	10	Purified locus E9	None
3	11	Purified locus E9	RE9
3	12	Purified locus E9	M3

6.5 Discussion

Our experiments show that good 3' agreement and overall good binding affinity are sufficient for priming success. For binding sites with affinity greater than -10 kilocalories per mole, products were obtained from purified PCR products containing the target template.

The cloning experiment shows that the thermodynamic models miss many real binding sites, as many of the clones mapped to sequences with no identifiable binding sites according to the thermodynamic models employed in this work. This is likely to be a result of inaccuracy in the model or the reference data used to compute binding energies.

Even though the model misses some binding sites, our approach can accurately design primers to amplify multiple targets. Thus, even though some parameters of the algorithm used to design primers to target multiple CpG sites in Chapter 5 may need some adjustment (i.e. the length of the 3' region of agreement and binding energy thresholds), the experiments in this chapter imply that a substantial fraction of primer pairs designed for CpG sites are likely to yield the desired PCR products.

Chapter 7

CONCLUSIONS AND FUTURE WORK

This work has presented an approach to PCR primer design based on DNA thermodynamics. The primer design method described works well, and is significantly simpler to use than competing approaches exemplified by the program Primer3.

Primer design methods must be assessed as to whether they can design primers that will amplify a target, and whether they can design primers that target specifically. The experimental results shown here support the claim the thermodynamic approach can design primers that amplify a target. However, evaluating specificity is harder; the data used to characterize the PCRs was not very informative for identifying nonspecific PCRs.

A very practical direction for future work would be to carefully determine rules to predict which primer binding sites will lead to extension by a polymerase enzyme. However, because some polymerases can extend primers even with 3' mismatches, this project would necessarily be polymerase specific. Several sets of rules would need to be developed for the common polymerases used in PCR reactions. However, identification of a standard reaction mixture for which PCR primer feasibility and specificity could be accurately predicted would be of significant benefit for the research community.

The method described for finding primer binding sites is likely to have applications in microarray probe design. Microarray hybridization experiments are much closer to thermodynamic equilibrium, and the binding site search will potentially improve predictions of probe specificity. Thus, another direction of future work would be to compare the binding sites for a probe to empirical measures of specificity. Such a research direction could signficantly improve microarray normalization.

The approach for multiple target primer design has not yet been systematically evaluated . Primers are currently being designed for a high throughput experiment that will measure methylation in CpG islands, but the empirical evaluation of the computational approach will be performed after this thesis is completed. The limited experimental results in bacteria indicate that although the expected products were obtained, so were a number of unexpected products; a better understanding of polymerase extension behavior will significantly improve this kind of primer design as well as the standard primer design problem.

Another important application is multiplex PCR, in which several fragments are simultaneously copied by a primer pool with several primer pairs. The methods in this work can be easily extended to deal with this kind of design problem; however, specificity will be much harder to address because of the quadratic increase in the number of primer pairs that could potentially participate in side reactions.

All of the computational methods developed in this work will be improved as chemists refine the models of DNA binding. Indeed, the field of DNA binding still has outstanding and fundamental questions. Perhaps the most relevant to this work is the degree to which the thermodynamic reference data exhibits temperature dependence. If there is a strong temperature dependence, then the reference data will have to be re-evaluated. However, any improvements in computation of binding affinities will directly benefit the methods described here, and so the ability of the models to predict PCR results will only improve.

In conclusion, I have made four contributions in this thesis. I have developed a method for rapid and sensitive search for primer binding sites in genomic databases; this is likely to be useful for PCR primer design, microarray probe design, and normalization of microarray data to account for probe specificity. I have developed a PCR primer design method based heavily on thermodynamic considerations, and showed that it works experimentally. This will be especially useful to experimentalists who need to design primers for the kinds of challenging regions for which Primer3 has difficulty. I have shown how this thermodynamic framework can be extended naturally to more complicated primer design problems, such as multiple target primer design. Finally, I have shown that DNA binding interactions can be usefully embedded in a kernel methods framework. In the multi-target application, this resulted in identification of sequences for which primers could be designed via standard kernel methods. This kernel approach to may be useful for further machine learning applications.

BIBLIOGRAPHY

[1] K. A. Abd-Elsalam. Bioinformatic tools and guidelines for PCR primer design. *African Journal of Biotechnology*, 2(5):91–95, 2003.

[2] E. S. Abrams, S. E. Murdaugh, and L. S. Lerman. Intramolecular DNA melting between stable helical segments: melting theory and metastable states. *Nucleic Acids Res.*, 23(14):2775–2783, 1995.

[3] N. Von Ahsen, C. T. Wittwer, and E. Schütz. Oligonucleotide melting temperatures under PCR conditions: Nearest-neighbor corrections for Mg^{2+}, Deoxynucleotide Triphosphate, and Dimethyl Sulfoxide concentrations with comparison to alternative empirical formulas. *Clinical Chemistry*, 32(12):1670–1693, 1992.

[4] H.T. Allawi and J. SantaLucia, Jr. Thermodynamics and NMR of internal GT mismatches in DNA. *Biochemistry*, 36(34):10581–10594, 1997.

[5] H.T. Allawi and J. SantaLucia, Jr. Nearest neighbor thermodynamic parameters for internal GA mismatches in DNA. *Biochemistry*, 37(8):2170–2179, 1998.

[6] H.T. Allawi and J. SantaLucia, Jr. Nearest-neighbor thermodynamics of internal AC mismatches in DNA: Sequence dependence and pH effects. *Biochemistry*, 37(26):9435–9444, 1998.

[7] H.T. Allawi and J. SantaLucia, Jr. Thermodynamics of internal CT mismatches in DNA. *Nucleic Acids Res.*, 26(11):2694–2701, 1998.

[8] S. F. Altschul, W. Gish, W. Miller, E. W. Myers, and D. J. Lipman. Basic local alignment search tool. 215:403–410, 1990.

[9] A. Andersson, R. Bernander, and P. Nilsson. Dual-genome primer design for construction of DNA microarrays. *Bioinformatics*, 21(3):325–332, 2005.

[10] M. A. Batzer and P. L. Deininger. Alu repeats and human genomic diversity. *Nat. Rev. Genetics*, 3:370–380, 2002.

[11] B.N. Belintsev, A.V. Vologodskii, and M.D. Frank-Kamenetskii. Influence of base sequence on the stability of the double helix of DNA. *Molekulyarnaya Biologiya*, 10(4):764–769, 1976.

[12] A. Ben-Hur. http://pyml.sourceforge.net.

[13] A. S. Benight and R. M. Wartell. Influence of base-pair changes and cooperativity parameters on the melting curves of short DNAs. *Biopolymers*, 22:1409–1425, 1983.

[14] A. S. Benight, R. M. Wartell, and D. K. Howell. Theory agrees with experimental thermal denaturation of short DNA restriction fragments. *Nature*, 289:203–205, 1981.

[15] D. P. Bertsekas. *Constrained Optimization and Lagrange Multiplier Methods*. Athena Scientific Press, Belmont, Massachusetts, 1982.

[16] R. Bhatia. Infinitely divisible matrices. *Am. Math. Monthly*, 113:221–235, 2006.

[17] R. D. Blake, J.W. Bizzaro, J.D. Blake, G.R. Day, S.G. Delcourt, J. Knowles, K.A. Marx, and J. SantaLucia, Jr. Statistical mechanical simulation of polymeric DNA melting with MELTSIM. *Bioinformatics*, 15:370–375, 1999.

[18] R. D. Blake and S. G. Delcourt. Thermodynamic effects of formamide on DNA stability. *Nucleic Acids Res.*, 24(11):2095–2103, 1996.

[19] R.D. Blake and S. G. Delcourt. Thermal stability of DNA. *Nucleic Acids Res.*, 26(14):3323–3332, 1998.

[20] S. Bommarito, N. Peyret, and J. SantaLucia, Jr. Thermodynamic parameters for DNA sequences with dangling ends. *Nucleic Acids Res.*, 28(9):1929–1934, 2000.

[21] D.F. Bradley, Jr. I. Tinoco, and R. W. Woody. Absorption and rotation of light by helical oligomers: the nearest neighbor approximation. *Biopolymers*, 1(3):239–267, 1963.

[22] K. J. Breslauer, R. Frank, H. Blocker, and L.A. Marky. Predicting DNA duplex stability from the base sequence. *Proc. Natl. Acad. Sci. U.S.A.*, 83(11):3746–3750, 1986.

[23] M. P. S. Brown, W. N. Grundy, D. Lin, N. Cristianini, C. W. Sugnet, T. S. Furey, M. Ares Jr., and D. Haussler. Knowledge-based analysis of microarray gene expression data by using support vector machines. *Proc. Natl. Acad. Sci. USA*, 97:262–267, 2000.

[24] J. Brownie, S. Shawcross, J. Theaker, D. Whitcombe, R. Ferrie, C. Newton, and S. Little. The elimination of primer-dimer accumulation in PCR. *Nucleic Acids Res.*, 25(16):3235–3241, 1997.

[25] A.F. Burden, N.C. Manley, A.D. Clark, S.M. Gartler, C.D. Laird, and R.S. Hansen. Hemimethylation and non-CpG methylation levels in a promoter region of human LINE-1 (L1) repeated elements. *J. Biol. Chem.*, (280):14413–14419, 2005.

[26] S.A. Bustin. Quantification of mRNA using real-time reverse transcription of PCR(RT-PCR): trends and problems. *J. Mol. Endocrin.*, 29:23–39, 2002.

[27] C.R. Cantor and P.R. Schimmel. *Biophysical Chemistry: Part III: The behavior of biological macromolecules*. W.H. Freeman and Company, New York, 1980.

[28] C.R. Cantor and C. L. Smith. *Genomics*. John Wiley and Sons, New York, 1999.

[29] J. B. Chaires. Possible origin of differences between van't Hoff and calorimetric enthalpy estimates. *Biophys. Chem.*, 64:15–23, 1997.

[30] R. Chakrabarti and C. E. Schutt. The enhancement of PCR amplification by low molecular weight amides. *Nucleic Acids Res.*, 29(11):2377–2381, 2001.

[31] T. Chalikian, J. Volker, G. Plum, and K. Breslauer. A more unified picture for the thermodynamics of nucleic acid duplex melting: A characterization by calorimetric and volumetric techniques. *Proc. Natl. Acad. Sci. USA*, 96:7853–7858, 2004.

[32] Chih-Chung Chang and Chih-Jen Lin. LIBSVM: a library for support vector machines. 2001.

[33] Q. Chou, M. Russell, D. E. Birch, J. Raymond, and W. Bloch. Prevention of pre-PCR mis-priming and primer dimerization improves low-copy-number amplifications. *Nucleic Acids Res.*, 20(7):1717–1723, 1992.

[34] S. J. Clark, J. Harrison, C. L. Paul, and M. Frommer. High sensitivity mapping of methylated cytosines. *Nucleic Acids Res.*, 22(15):2990–2997, 1994.

[35] M. Collins and R. M. Myers. Alterations in DNA helix stability due to base modifications can be evaluated using denaturing gradient gel electrophoresis. *J. Mol. Bio.*, 198:737–744, 1987.

[36] A. Cooper. Heat capacity of hydrogen-bonded networks: an alternative view of protein folding thermodynamics. *Biophys. Chem.*, 85:25–39, 2000.

[37] A. Cooper, C. M. Johnson, J. H. Lakey, and M. Nöllmann. Heat does not come in different colours:entropy-enthalpy compensation, free energy windows, quantum confinement, pressure perturbation calorimetry, solvation, and the

multiple causes of heat capacity effects in biomolecular interactions. *Biophys. Chem.*, 93:215–230, 2001.

[38] C. Cortes, M. Mohri, and J. Weston. A general regression technique for learning transductions. In *Proc. 22nd Int. Conf. on Machine Learning*, Bonn, Germany, August 7-11 2005.

[39] G. Cosa, K.-S. Focseanu, J.R.N. McLean, J. P. McNamee, and J.C. Scaiano. Photophysical properties of fluorescent DNA-dyes bound to single and double stranded DNA in aqueous buffred solution. *Photochem. and PhotoBiol.*, 73(6):585–599, 2001.

[40] N. Cristianini and J. Shawe-Taylor. *An Introduction to Support Vector Machines and Other Kernel Based Learning Methods*. Cambridge University Press, Cambridge, UK, 2000.

[41] D.M. Crothers. Calculation of melting curves for DNA. *Biopolymers*, 6(10):1391–1404, 1968.

[42] D.M. Crothers and N.R. Kallenbach. On the helix-coil transition in heterogeneous polymers. *J. Chem. Phys*, 45(3):917–927, 1966.

[43] R. A. J. Darby, M. Sollogoub, C. McKeen, L. Brown, A. Risitano, N. Brown, C. Barton, T. Brown, and K. R. Fox. High throughput measurement of duplex, triplex and quadruplex melting curves using molecular beacons and a LightCycler. *Nucleic Acids Res.*, 30(9):e39, 2002.

[44] S. G. Delcourt and R. D. Blake. Stacking energies in DNA. *J. Biol. Chem.*, 266(23):15160–15169, 1991.

[45] R. A. Dimitrov and M. Zuker. Prediction of hybridization and melting for double stranded nucleic acids. *Biophys. Journal*, 87(1):215–226, 2004.

[46] R. M. Dirks and N. A. Pierce. A partition function algorithm for nucleic acid secondary structure including pseudoknots. *Journal of Computational Chemistry*, 24(13):1664–1677, 2003.

[47] M.J. Doktycz, R. F. Goldstein, T. M. Paner, F. J. Gallo, and A. S. Benight. Studies of DNA dumbbells. I. melting curves of 17 DNA dumbbells with different duplex stem sequences linked by T_4 endloops: Evaluation of the nearest-neighbor stacking interactions in DNA. *Biopolymers*, 32(7):849–864, 1992.

[48] M. O. Dorschner, M. Hawrylycz, R. Humbert, J. C. Wallace, A. Shafer, J. Kawamoto, J. Mack, R. Hall, J. Goldy, P. J. Sabo, A. Kohli, Q. Li, M. McArthur, and J. Stamatoyannopoulos. High-throughput localization of

functional elements by quantitative chromatin profiling. *Nat. Methods*, 1:219–225, 2004.

[49] R.O. Duda, P.E. Hart, and D.G. Stork. *Pattern Classification (2nd Edition)*. Wiley-Interscience, 2000.

[50] ENCODE Project Consortium. The ENCODE (ENCyclopedia of DNA elements) project. *Science*, 306:636–640, 2004.

[51] D. Erie, N. Sinha, W. Olson, R. Jones, and K. Breslauer. A dumbbell-shaped, double hairpin structure of DNA: A thermodynamic investigation. *Biochemistry*, 26:7150–7159, 1987.

[52] S. Essono, Y. Frobert, J. Grassiand C. Creminon, and D. Boquet. A general method allowing the design of oligonucleotide primers to amplify the variable regions from immunoglobulin cDNA. *Journal of Immunological Methods*, 279(1-2):251–266, 2003.

[53] R. J. Fernandes and S. S. Skiena. Microarray synthesis through multiple-use PCR primer design. *Bioinformatics*, 1(1):1–8, 2002.

[54] R. F.Goldstein and A. S. Benight. How many numbers are required to specify sequence dependent properties of polynucleotides? *Biopolymers*, 32(12):1670–1693, 1992.

[55] R. Fislage, M. Berceanu, Y. Humboldt, M. Wendt, and H. Oberender. Primer design for a prokaryotic differential display RT-PCR. *Nucleic Acids Res.*, 25(9):1830–1835, 1997.

[56] M. Fixman and J. J. Freire. Theory of DNA melting curves. *Biopolymers*, 16(12):2693–2704, 1977.

[57] J. Fredslund, L. Schauser, L.H. Madsen, N. Sandal, and J. Stougaard. PriFi: using a multiple alignment of related sequences to find primers for amplification of homologs. *Nucleic Acids Res.*, 33:W516–W520, 2005.

[58] M. Frommer, L.E. McDonald, D.S. Millar, C.M. Collis, F. Watt, G.W. Grigg, P.L. Molloy, and C.L. Paul. A genomic sequencing protocol that yields a positive display of 5-methylcytosine residues in individual DNA strands. *Proc. Natl. Acad. Sci. USA*, 89:1827–1831, 1992.

[59] X. Gao, P. Yo, A. Keith, T. J. Ragan, and T. K. Harris. Thermodynamically balanced inside-out (TBIO) PCR-based gene synthesis: a novel method of primer design for high-fidelity assembly of longer gene sequences. *Nucleic Acids Res.*, 31(22):e143, 2002.

[60] T. Garel and H. Orland. Generalized Poland-Scheraga model for DNA hybridization. *Biopolymers*, 75(6):453–467, 2004.

[61] P. M. K. Gordon and C. W. Sensen. Osprey: a comprehensive tool employing novel methods for the design of oligonucleotides for DNA sequencing and microarrays. *Nucleic Acids Res.*, 32(17):e133, 2004.

[62] O. Gotoh and Y. Tagashira. Stabilities of nearest-neighbor doublets in double-helical DNA determined by fitting calculated melting profiles to observed profiles. *Biopolymers*, 20:1033–1042, 1981.

[63] D. M. Gray. Derivation of nearest-neighbor properties from data on nucleic acid oligomers. I. simple sets of independent sequences and the influence of absent nearest neighbors. *Biopolymers*, 42(7):783–793, 1997.

[64] D. M. Gray. Derivation of nearest-neighbor properties from data on nucleic acid oligomers. II. thermodynamic parameters of DNA.RNA hybrids and DNA duplexes. *Biopolymers*, 42(7):795–810, 1997.

[65] D. M. Gray and Jr. I. Tinoco. A new approach to the study of sequence-dependent properties of polynucleotides. *Biopolymers*, 9(2):223–244, 1970.

[66] R. Griffais, P.M. Andre, and M. Thibon. K-tuple frequency in the human genome and polymerase chain reaction. *Nucleic Acids Res.*, 19(14):3887–3891, 1991.

[67] D. Gusfield. *Algorithms on Strings, Trees, and Sequences*. Cambridge University Press, Cambridge, UK, 1997.

[68] S. Haas, M. Vingron, A. Poustka, and S. Wiemann. Primer design for large scale sequencing. *Nucleic Acids Res.*, 26(12):3006–3012, 1998.

[69] S. A. Haas, M. Hild, A. P. H. Wright, T. Hain, D. Talibi, and M. Vingron. Genome-scale design of PCR primers and long oligomers for DNA microarrays. *Nucleic Acids Res.*, 31(19):5576–5581, 2003.

[70] W.P. Halford, V.C. Falco, B.M. Gebhardt, and D.J. Carr. The inherent quantitative capacity of the reverse transcription-polymerase chain reaction. *Anal. Biochem.*, 266(2):181–191, 1999.

[71] R.P. Haugland. *Handbook of Fluorescent Probes and Research Chemicals*. Molecular Probes Inc., Eugene, OR, 2001.

[72] R. Higuchi, G. Dollinger, P.S. Walsh, and R. Griffith. Simultaneous amplification and detection of specific DNA-sequences. *Biotechnology*, 10(4):413–417, 1992.

[73] L. Hillier and P. Green. OSP: A computer program for choosing PCR and DNA sequencing primers. *PCR Meth. Appl.*, 1(2):124–128, 1991.

[74] J.A. Holbrook, M. W. Capp, R. M. Saecker, and M. Thomas Record Jr. Enthalpy and heat capacity changes for formation of an oligomeric DNA duplex: interpretation in terms of coupled processes of formation and association of single-stranded helices. *Biochemistry*, 38:8409–8422, 1999.

[75] J. Hong, M. W. Capp, C. F. Anderson, R. M. Saecker, D. J. Felitsky, M. W. Anderson, and M. T. Record. Preferential interactions of glycine betaine and of urea with DNA: implications for DNA hydration and for effects of these solutes on DNA stability. *Biochemistry*, 43:14744–14758, 2004.

[76] R. A. Horn and C. R. Johnson. *Matrix Analysis*. Cambridge University Press, 1985.

[77] R. A. Horn and C. R. Johnson. *Topics in Matrix Analysis*. Cambridge University Press, 1991.

[78] M-H Hsieh, W-C Hsu, S-K Chiu, and C-M Tzeng. An efficient algorithm for minimal primer set selection. *Bioinformatics*, 19(2):285–286, 2003.

[79] C. G. Huber and H. Oberacher. Analysis of nucleic acids by on-line liquid chromatography mass spectrometry. *Mass Spectrometry Reviews*, 20(5):310–343, 2001.

[80] G. B. Hurst, K. Weaver, M. J. Doktycz, M. V. Buchanan, A. M. Costello, and M. E. Lidstrom. MALDI-TOF analysis of Polymerase Chain Reaction products from methanotrophic bacteria. *Anal. Chem.*, 70(13):2693–2698, 1998.

[81] D. Hyndman, A. Cooper, S. Pruzinsky, D. Coad, and M. Misuhashi. Software to determine optimal oligonucleotide sequences based on hybridization simulation data. *BioTechniques*, 20(6):1090–1097, 1996.

[82] M. A. Innis, D. H. Gelfand, and J. J. Sninsky. *PCR Applications: Protocols for Functional Genomics*. Academic Press, 1999.

[83] Jr. J. SantaLucia and D. Hicks. The thermodynamics of DNA structural motifs. *Annu. Rev. Biomol. Struct.*, 33:415–440, 2004.

[84] S.N. Jarman. Amplicon: software for designing PCR primers on aligned DNA sequences. *Bioinformatics*, 20(10):1644–1645, 2004.

[85] S. J. Johnson and L. S. Beese. Structures of mismatch replication errors observed in a DNA polymerase. *Cell*, 116:803–816, 2004.

[86] B. Kaltenboeck and C.M. Wang. Advances in real-time PCR: Application to clinical laboratory diagnostics. *Adv. in Clin. Chem.*, 40:219–259, 2005.

[87] T. Kämpke, M. Kieninger, and M. Mecklenburg. Efficient primer design algorithms. *Bioinformatics*, 17(3):214–225, 2001.

[88] H. J. Karlsson, M. Eriksson, E. Perzon, B. Akerman, P. Lincoln, and G. Westman. Groove-binding unsymmetrical cyanine dyes for staining of DNA: syntheses and characterization of the DNA-binding. *Nucleic Acids Res.*, 31(21):6227–6234, 2003.

[89] S.-H. Ke and R. M. Wartell. The thermal stablity of DNA fragments with tandem mismatches at a d(CXYG)-d(CY'X'G) site. *Nucleic Acids Res.*, 25(4):707–712, 1996.

[90] W.J. Kent, C. W . Sugnet, T.S. Furey, K.M. Roskin, T.H. Pringle, A.M. Zahler, and D. Haussler. The Human Genome Browser at UCSC. *Genome Res.*, 12(6):996–1006, 2002.

[91] S. Kirkpatrick, C.D. Gelatt, and M.P. Vecchi. Optimization by simulated annealing. *Science*, 220(4598):671–680, 1983.

[92] H.H. Klump. Energetics of order/order transitions in nucleic acids. *Can. J. Chem.*, 66:804–811, 1988.

[93] M.T. Krahmer, J.J. Walters, K.F. Fox, A. Fox, K.E. Creek, L. Pirisi, D. S. Wunschel, R. D. Smith, D. L. Tabb, and III J. R. Yates. MS for identification of single nucleotide polymorphisms and MS/MS for discrimination of isomeric PCR products. *Anal. Chem.*, 72(17):4033–4040, 2000.

[94] S. Kwok, D. E. Kellogg, N. McKinney, D. Spasic, L. Goda, C. Levenson, and J.J. Snisky. Effects of primer-template mismatches on the polymerase chain reaction: Human immunodeficiency virus type 1 model studies. *Nucleic Acids Res.*, 18(1):999–1005, 1990.

[95] A. N. Lane and T. C. Jenkins. Thermodynamics of nucleic acids and their interactions with ligands. *Quart. Rev. Biophys*, 33(3):255–306, 2000.

[96] A. Larionov, A. Krause, and W. Miller. A standard curve based method for relative real time pcr data processing. *BMC Bioinformatics*, 6:62, 2005.

[97] I.N. Levine. *Physical Chemistry*. McGraw Hill, Boston, 2003.

[98] F. Li and G.D. Stormo. Selection of optimal DNA oligos for gene expression arrays. *Bioinformatics*, 17(11):1067–1076, 2001.

[99] L-C. Li and R. Dahiya. Methprimer: designing primers for methylation PCRs. *Bioinformatics*, 18(11):1427–1431, 2002.

[100] P. Li, K. C. Kupfer, C.J. Davies, D. Burbee, G. A. Evans, and H. R. Garner. PRIMO: A primer design program that applies base quality statistics for automated large-scale DNA sequencing. *Genomics*, 40(3):476–485, 1997.

[101] W. Li, B. Xi, W. Yang, M. Hawkins, and U. K. Schubart. Complex DNA melting profiles of small PCR products revealed using SYBR Green I. *BioTechniques*, 35(4):702–706, 2003.

[102] C. Linhart and R. Shamir. The degenerate primer design problem. *Bioinformatics*, 18(Suppl. 1):S172–S180, 2002.

[103] W. Liu and D. A. Saint. A new quantitative method of real time reverse transcription polymerase chain reaciton assay based on simulation of polymerase chain reaction kinetics. *Anal. Bioc.*, 302:52–59, 2002.

[104] W. Liu and D. A. Saint. Validation of a quantitative method of real time PCR kinetics. *Bioc. and Bioph. Res. Comm.*, 294:347–353, 2002.

[105] C. E. Lopez-Nieto and S. K. Nigam. Selective amplification of protein-coding regions of large sets of genes using statistically designed primer sets. *Nat. Biotechnol.*, 14(7):857–861, 1996.

[106] T. Lowe, J. Sharefkin, S. Q. Yang, and C. W. Dieffenbach. A computer program for selection of oligonucleotide primers for polymerase chain reactions. *Nucleic Acids Res.*, 18(7):1757–1761, 1990.

[107] L.S.Lerman and K. Silverstein. Computational simulation of DNA melting and its application to denaturing gradient gel electrophoresis. *Meth. Enzym.*, 155:482–501, 1987.

[108] U. Manber and E. Myers. Suffix arrays: a new method for on-line serach. *SIAM J. Comput.*, 2:935–948, 1993.

[109] T.P. Mann, R. Humbert, J. A. Stamatoyannopolous, and W.S. Noble. Automated validation of polymerase chain reactions using amplicon melting curves. In *CSB2005 Computational Systems Bioinformatics*, Stanford,California, August 8-11 2005.

[110] T.P. Mann and W.S. Noble. Efficient identification of DNA binding partners in a sequence database. *Bioinformatics*, 22(14):e350–e358, 2006.

[111] N.R. Markham and M. Zuker. DINAMelt web server for nucleic acid melting prediction. *Nucleic Acids Res.*, 33:W577–W581, 2005.

[112] D.W. Marquardt. An algorithm for least-squares estimation for non-linear parameters. *J. Soc. Ind. Appl. Math.*, 11:431–441, 1963.

[113] D. H. Mathews, J. Sabina, M. Zuker, and D. H. Turner. Expanded sequence dependence of thermodynamic parameters improves prediction of RNA secondary structure. *J. Mol. Biol.*, 288:911–940, 1999.

[114] C. M. McCallum, L. Comai, E.A. Greene, and S. Henikoff. Targeting induced local lesions in genomes (TILLING) for plant functional genomics. *Plant Physiology*, 123(2):439–442, 2000.

[115] J. S. McCaskill. The equilibrium partition function and base pair binding probabilities for RNA secondary structure. *Biopolymers*, 29(6–7):1105–1119, 1988.

[116] J. A. McDowell and D. H. Turner. Investigation of structural basis for thermodynamic stabilities of tandem GU mismatches: solution structure of (rGAGGUCUC))2 by two-dimensional NMR and simulated annealing. *Biochemistry*, 34:14077–14089, 1996.

[117] S. J. McKay and S. J. M. Jones. AcePrimer: automation of PCR primer design based on gene structure. *Bioinformatics*, 18(11):1538–1539, 2002.

[118] M. W. Mecklenburg. Design of high annealing temperature PCR primers and their use in the development of a versatile low-copy-number amplification protocol. *Adv. Mol. and Cell Bio.*, 15b:473–490, 1996.

[119] G. Mei and S. H. Hardin. Octamer-primed sequencing technology: development of primer identification software. *Nucleic Acids Res.*, 7:e22, 2000.

[120] C. E. Metz. Basic principles of ROC analysis. *Semin. Nucl. Med.*, 8:283–298, 1978.

[121] P.J. Mikulecky and A. L. Feig. Heat capacity changes associated with DNA duplex formation: Salt-and sequence-dependent effects. *Biochemistry*, 45:614–616, 2006.

[122] P.J. Mikulecky and A. L. Feig. Heat capacity changes associated with nucleic acid folding. *Biopolymers*, 82:38–58, 2006.

[123] B. E. Miner, R. J. Stöger, A. F. Burden, C. D. Laird, and R. S. Hansen. Molecular barcodes detect redundancy and contamination in hairpin-bisulfite PCR. *Nucleic Acids Res.*, 32(17):e135, 2004.

[124] F. Miura, C. Uematsu, Y. Sakaki, and T. Ito. A novel strategy to design highly specific PCR primers based on the stability and uniqueness of the 3'-end subsequences. *Bioinformatics*, 21(24):4363–4370, 2005.

[125] L. Movileanu, J. M. Benevides, and G. J. Thomas Jr. Determination of base and backbone contributions to the thermodynamics of premelting and melting transitions in B DNA. *Nucleic Acids Res.*, 30(17):3767–3777, 2002.

[126] E. Myers. A sublinear algorithm for approximate keyword matching. *Algorithmica*, 12(4-5):345–374, 1994.

[127] R. M. Myers, S. J. Fischer, T. Maniatis, and L. S. Lerman. Modification of the melting properties of duplex DNA by attachment of a GC-rich DNA sequence as determined by denaturing gradient gel electrophoresis. *Nucleic Acids Res.*, 13(9):3111–3129, 1985.

[128] R. M. Myers, L. S. Lerman, and T. Maniatis. A general method for saturation mutagenesis of cloned DNA fragments. *Science*, 229(4710):242–247, 1985.

[129] R. M. Myers, T. Maniatis, and L. S. Lerman. Detection and localization of single base changes by denaturing gradient gel electrophoresis. *Meth. Enzym.*, 155:501–527, 1985.

[130] A. W. Naylor and G. R. Sell. *Linear Operator Theory in Engineering and Science.* Springer, 1982.

[131] M. M. Neff, J. D. Neff, J. Chory, and A. E. Pepper. dCAPS, a simple technique for the genetic analysis of single nucleotide polymorphisms:experimental approaches in *arabidopsis thaliana* genetics. *The Plant Journal*, 14(3):387–392, 1998.

[132] J.A. Nelder and R. Mead. A simplex method for function minimization. *Computer Journal*, 7:308–313, 1965.

[133] C. R. Newton, A. Graham, L.E. Heptinstall, S.J.Powell, C.Summers, N. Kalsheker, J.C. Smith, and A. F. Markham. Analysis of any point mutation in DNA. the amplification refractory mutation system(ARMS). *Nucleic Acids Res.*, 17(7):2503–2516, 1989.

[134] W.S. Noble. Support vector machine applications in computational biology. In B. Schölkopf, K. Tsuda, and J.-P. Vert, editors, *Kernel Methods in Computational Biology*. MIT Press, Cambridge, MA, USA, 2004.

[135] R. Owczarzy, I. Dunietz, M. Behlke, I. M. Klotz, and J. A. Walder. Thermodynamic treatment of oligonucleotide duplex-simplex equilibria. *Proc. Natl. Acad. Sci. USA*, 100(25):14840–14845, 2003.

[136] R. Owczarzy, P. M. Vallone, F. J. Gallo, T. M. Paner, M. J. Lane, and A. S. Benight. Predicting sequence-dependent melting stability of short duplex DNA oligomers. *Biopolymers*, 44(3):217–239, 1997.

[137] R. Owczarzy, Y. You, B. G. Moreira, J. A. Manthey, L. Huang, M. A. Behlke, and J. A . Walder. Effects of sodium ions on DNA duplex oligomers: Improved predictions of melting temperatures. *Biochemistry*, 43(12):3537–3554, 2004.

[138] F. Pattyn, F. Speleman, A. De Paepe, and J. Vandesompele. RTPrimerDB: the real-time PCR primer and probe database. *Nucleic Acids Res.*, 31(1):122–123, 2003.

[139] W. Pearson, G. Robins, D. Wrege, and T. Zhang. New approach to primer selection in polymerase chain reaction experiments. In *Proc. Intl. Conf. on Intelligent Systems for Molecular Biology*, Cambridge, England, July 1995.

[140] S. N. Peirson, J. N. Butler, and R. G. Foster. Experimental validation of novel and conventional approaches to quantitative real-time PCR data analysis. *Nucleic Acids Res.*, 31(14):e73, 2003.

[141] M.P. Perelroyzen, V.I. Lyamichev, Y. Kalambet, Y. Lyubchenko, and A.V. Vologodskii. A study of the reversibility of helix-coil transitions in DNA. *Nucleic Acids Res.*, 9(16):4043–4059, 1981.

[142] M. Petersheim and D. H. Turner. Base-stacking and base-pairing contributions to helix stability: Thermodyanmics of double-helix formation with CCGG,CCGGp,CCGAp,ACCGGp,CCGGUp, and ACCGGUp. *Biochemistry*, 22:256–263, 1983.

[143] N. Peyret, P. A. Seneviratne, H.T. Allawi, and J. SantaLucia, Jr. Nearest-neighbor thermodynamics and NMR of DNA sequences with internal AA, CC, GG, and TT mismatches. *Biochemistry*, 38(12):3468–3477, 1999.

[144] J. Platt. Probabilistic outputs for support vector machines and comparisons to regularized likelihood methods. In P.J. Bartlett, B. Schölkopf, D. Schuurmans, and A.J. Smola, editors, *Advances in Large Margin Classifiers*. MIT Press, Cambridge, MA, USA, 1999.

[145] R. M. Podowski and E. L. L. Sonnhammer. MEDUSA: large scale automatic selection and visual assessment of PCR primer pairs. *Bioinformatics*, 17(7):656–657, 2001.

[146] D. Poland. Recursion relation generation of probability profiles for specific-sequence macromolecules with long-range correlations. *Biopolymers*, 13(9):1859–1871, 1974.

[147] D. Poland. DNA melting profiles from a matrix method. *Biopolymers*, 73:216–228, 2004.

[148] A. L. Price, E. Eskin, and P. A. Pevzner. Whole-genome analysis of Alu repeat elements reveals complex evolutionary history. *Genome Res.*, 14:2245–2252, 2004.

[149] D. Proudnikov, V. Yuferov, Y. Zhou, K. S. LaForge, A. Ho, and M. J. Kreek. Optimizing primer-probe design for fluorescent PCR. *Journal of Neuroscience Methods*, 123(1):31–45, 2003.

[150] H.J. Purohit, D. V. Raje, and A. Kapley. Identification of signature and primers specific to genus pseuodomonas using mismatched patterns of 16s rDNA sequences. *BMC Bioinformatics*, 4:19, 2003.

[151] R. S. Quartin and J. G. Wetmur. Effect of ionic strength on the hybridization of oligonucleotides with reduced charge due to methylphosphonate linkages to unmodified oligodeoxynucleotides containing the complementary sequence. *Biochemistry*, 28:1040–1047, 1989.

[152] G. Raddatz, M. Dehio, T. F. Meyer, and C. Dehio. Primearray: genome-scale primer design for DNA-microarray construction. *Bioinformatics*, 17(1):98–99, 2001.

[153] G. Ratsch, S. Sonnenburg, J. Srinivasan, H. Witte, K.R. Muller, R. Sommer, and B. Schölkopf. Improving the *c. elegans* genome annotation using machine learning. *PLoS Computational Biology*, (2):e20, 2007.

[154] A. Razin. CpG methylation, chromatin structure and gene silencing-a three-way connection. *EMBO J.*, 17(17):4905–4908, 1998.

[155] William A. Rees, Thomas D. Yager, John Korte, and Peter H. Von Hippel. Betaine can eliminate the base pair composition dependence of DNA melting. *Biochemistry*, 32(1):137–144, 1993.

[156] C. Regeard, J. Maillard, and C. Holliger. Development of degenerate and specific PCR primers for the detection and isolation of known and putative chloroethene reductive dehalogenase genes. *Journal of Microbiological Methods*, 56(1):107–118, 2004.

[157] D. Rentzeperis, J. Ho, and L. A. Marky. Contribution of loops and nicks to the formation of DNA dumbbells: Melting behavior and ligand binding. *Biochemistry*, 32:2564–2572, 1993.

[158] S.A. Rice and A. Wada. On a model of the helix-coil transition in macromolecules. II. *J. Chem. Phys*, 29:233, 1958.

[159] F. S. Roberts. *Applied Combinatorics*. Prentice-Hall, Inc., Eaglewood Cliffs, New Jersey, 1984.

[160] T. M. Rose, J. G. Henikoff, and S. Henikoff. CODEHOP (COnsensus-DEgenerate Hybrid Oligonucleotide Primer) PCR primer design. *Nucleic Acids Res.*, 31(13):3763–3766, 2003.

[161] T. M. Rose, E. R. Schultz, J. G. Henikoff, S. Pietrokovski, C. M McCallum, and S. Henikoff. Consensus-degenerate hybrid oligonucleotide primers for amplification of distantly related sequences. *Nucleic Acids Res.*, 26(7):1628–1635, 1998.

[162] E. C. Rouchka, A. Khalyfa, and N.G.F. Cooper. Selection of oligonucleotide probes for protein coding sequences. *BMC Bioinformatics*, 6(175), 2005.

[163] I. Rouzina and V. A. Bloomfeld. Heat capacity effects on the melting of DNA. 1. general aspects. *Biophys. J.*, 77(6):3242–3255, 1999.

[164] S. Rozen and H. Skaletsky. Primer3 on the WWW for general users and for biologist programmers. In S. A. Krawetz and S. Misener, editors, *Bioinformatics Methods and Protocols: Methods in Molecular Biology*, pages 365–386. Humana Press, Totowa, NJ, 2000.

[165] E. Rubin and A. A. Levy. A mathematical model and a computerized simulation of PCR using complex templates. *Nucleic Acids Res.*, 24(18):3538–3545, 1996.

[166] W. Rychlik. Selection of primers for polymerase chain reaction. *Mol. Biotechnol.*, 3(2):129–134, 1995.

[167] W. Rychlik and R. E. Rhoads. A computer program for choosing optimal oligonucleotides for filter hybridization, sequencing, and in vitro amplification of DNA. *Nucleic Acids Res.*, 17(21):8543–8551, 1989.

[168] P. J. Sabo, M. Hawrylycz, J. C. Wallace, R. Humbert, M. Yu, A. Schafer, J. Kawamoto, R. Hall, J. Mack, M. O. Dorschner, M. McArthur, and J. A. Stamatoyannopoulos. Discovery of functional noncoding elements by digital analysis of chromatin structure. *Proc. Natl. Acad. Sci. U.S.A.*, 101(48):16837–16842, 2004.

[169] P. J. Sabo, R. Humbert, M. Hawrylycz, J. C. Wallace, M. O. Dorschner, M. McArthur, and J. A. Stamatoyannopoulos. Genome-wide identification of DNaseI hypersensitive sites using active chromatin sequence libraries. *Proc. Natl. Acad. Sci. U.S.A.*, 101(13):4537–4542, 2004.

[170] R. K. Saiki, D. H. Gelfand, S. Stoffel, S.J. Scharf, R. Higuchi, G. T. Horn, K. B. Mullis, and H.A.Erlich. Primer-directed enzymatic amplification of DNA with a thermostable polymerase. *Science*, 239(4839):487–491, 1988.

[171] J. SantaLucia, Jr. A unified view of polymer, dumbbell, and oligonucleotide DNA nearest-neighbor thermodynamics. *Proc. Natl. Acad. Sci. USA*, 95:1460–1465, 1998.

[172] J. SantaLucia Jr. Physical principles and Visual-OMP software for optimal PCR design. In A. Yuryev, editor, *Methods in Molecular Biology:PCR Primer Design*. Humana Press, Totowa, New Jersey, 2006.

[173] J. SantaLucia, Jr, H.T. Allawi, and P. A. Seneviratne. Improved nearest-neighbor parameters for predicting DNA duplex stability. *Biochemistry*, 35(11):3555–3562, 1996.

[174] J. SantaLucia, Jr and D. Hicks. The thermodynamics of DNA structural motifs. *Annu. Rev. Biophys. Biomol. Struct.*, 33:415–440, 2004.

[175] M. Schena, D. Shalon, R. W. Davis, and P. O. Brown. Quantitative monitoring of gene expression patterns with a complementary DNA microarray. 270:467–470, 1995.

[176] T.D. Schmittgen, B.A. Zakrajsek, A.G. Mills, V. Gorn, M.J. Singer, and Reed MW. Quantitative reverse transcription-polymerase chain reaction to study mRNA decay: comparison of endpoint and real-time methods. *Anal. Biochem.*, 285(2):194–204, 2000.

[177] B. Schölkopf, J.C. Platt, J. Shawe Taylor, A. J. Smola, and R. C. Williamson. Estimating the support of a high dimensional distribution. *Neural Computation*, 13(7):1443–1471, 2001.

[178] G. D. Schuler. Sequence mapping by electronic PCR. *Genome Res.*, 7(5):541–550, 1997.

[179] C. Schweitzer and J. C. Scaiano. Selective binding and local photophysics of the fluorescent cyanine dye PicoGreen in double-stranded and single-stranded DNA. *Phys. Chem. Chem. Phys.*, 5:4911–4917, 2003.

[180] S.E.Whitney, A. Sudhir, R. M. Nelson, and H. J. Viljoen. Principles of rapid polymerase chain reactions: mathematical modeling and experimental verification. *Comp. Bio. and Chem.*, 28:195–209, 2004.

[181] K. Sharp. Entropy-enthalpy compensation: fact or artifact? *Protein Science*, 10:661–667, 2001.

[182] J. Shawe-Taylor and N. Cristianini. *Kernel Methods for Pattern Analysis*. Cambridge University Press, Cambridge, UK, 2004.

[183] M. Simsek and H. Adnan. Effect of single mismatches at 3' end of primers on polymerase chain reaction. *Medical Sciences*, 2:11–14, 2000.

[184] A.F.A. Smit, R. Hubley, and P. Green. Repeatmasker open-3.0, http://www.repeatmasker.org, 1996–2004.

[185] W. R. Smith and R. W. Missen. *Chemical reaction equilibrium analysis : theory and algorithms*. Wiley, New York, 1982.

[186] G. Steger. Thermal denaturation of double-stranded nucleic acids: prediction of temperatures critical for gradient gel electrophoresis and polymerase chain reaction. *Nucleic Acids Res.*, 22(14):2760–2768, 1994.

[187] R. Stöger. In vivo methylation patterns of the leptin promoter in human and mouse. *Epigenetics*, 4(1):155–162, 2006.

[188] R. B. Stoughton. Applications of DNA microarrays in biology. *Annual Rev. Biochem.*, 74:53–82, 2005.

[189] N. Sugimoto, K. Honda, and M. Sasaki. Application of the thermodynamic parameters of DNA stability prediction to double-helix formation of deoxyribooligonucleotides. *Nucleosides & Nucleotides*, 13(6&7):1311–1317, 1994.

[190] N. Sugimoto, S. Nakano, M. Yoneyama, and K. Honda. Improved thermodynamic parameters and helix initiation factor to predict stability of DNA duplexes. *Nucleic Acids Res.*, 24(22):4501–4505, 1996.

[191] S. Swillens, J.-C. Goffard, Y. Marechal, A. de K. d'Exaerde, and H. E. Housni. Instant evaluation of the absolute initial number of cDNA copies from a single real-time PCR curve. *Nucleic Acids Res.*, 32(6):e53, 2004.

[192] V. Thareau, P. Dehais, C. Serizet, P. Hilson, P. Rouze, and S. Aubourg. Automatic design of gene-specific sequence tags for genome-wide functional studies. *Bioinformatics*, 19(17):2191–2198, 2003.

[193] A. Tichopad, M. Dilger, G. Schwarz, and M. W. Pfaffl. Standardized determinaton of real-time PCR efficiency from a single reaction set-up. *Nucleic Acids Res.*, 31(20):e122, 2003.

[194] A. Tikhomirova, I. V. Beletskaya, and T. V. Chalikian. Stability of DNA duplexes containing GG,CC,AA, and TT mismatches. *Biochemistry*, 45:10563–10571, 2006.

[195] A. Tikhomirova, N. Taulier, and T. V. Chalikian. Energetics of nucleic acid stability. *J. Am. Chem. Soc*, 126:16387–16394, 2004.

[196] K. R. Tindall and T. A. Kunkel. Fidelity of DNA synthesis by the thermus aquaticus DNA polymerase. *Biochemistry*, 27(16):6008–6013, 1988.

[197] E. Tostesen, F. Liu, T.-K. Jenssen, and E. Hovig. Speed-up of DNA melting algorithm with complete nearest neighbor properties. *Biopolymers*, 70:364–376, 2003.

[198] V. Vapnik. *Statistical Learning Theory*. John Wiley and Sons, New York, 1998.

[199] K. Varadaraj and D. M. Skinner. Denaturants or cosolvents improve the specificity of PCR amplification of a G+C rich DNA using genetically engineered DNA polymerases. *Gene*, 140(1):1–5, 1994.

[200] C. Varotto, E. Richly, F. Salamini, and D. Leister. GST-PRIME: a genome-wide primer design software for the generation of gene sequence tags. *Nucleic Acids Res.*, 29(21):4373–4377, 2001.

[201] M. V. Velikanov and R. Kapral. Polymerase chain reaction:a markov process approach. *J. Theor. Biol.*, 201:239–249, 1999.

[202] J. Volker, R. D. Blake, S. G. Delcourt, and K. J. Breslauer. High-resolution calorimetric and optical melting profiles of DNA plasmids: resolving contributions from intrinsic melting domains and specifically designed inserts. *Biopolymers*, 50:303–318, 1999.

[203] A.V. Vologodskii, B. R. Amirikyan, Y. Lyubchenko, and M.D. Frank-Kamenetskii. Allowance for heterogeneous stacking in the DNA helix-coil transition theory. *J. Biomol. Struct. and Dyn.*, 2(1):131–148, 1984.

[204] H.L. Vu, S. Troubetzkoy, H.H. Nguyen, M.W. Russell, and J. Mestecky. A method for quantification of absolute amounts of nucleic acids by (RT)-PCR and a new mathematical model for data analysis. *Nucleic Acids Res.*, 28(7):E18, 2000.

[205] A. Wada and A. Suyama. Local stability of DNA and RNA secondary structure and its relation to biological functions. *Prog. Biophys. Molec. Biol.*, 47(2):113–157, 1986.

[206] J.-Y. Wang and K. Drlica. Modeling hybridization kinetics. *Mathematical Biosciences*, 183:37–47, 2003.

[207] R. Y.-H. Wang, C. W. Gehrke, and M. Ehrlich. Comparison of bisulfite modification of 5-methyldeoxycytidine and deoxycitidine residues. *Nucleic Acids Res.*, 20(8):4777–4790, 1980.

[208] W. Wang and D. H. Johnson. Computing linear transforms of symbolic sequences. *IEEE Trans. Sig. Proc.*, 50(3):628–634, 2002.

[209] X. Wang and B. Seed. A PCR primer bank for quantitative gene expression analysis. *Nucleic Acids Res.*, 31(24):e154, 2003.

[210] X. Wang and B. Seed. Selection of oligonucleotide probes for protein coding sequences. *Bioinformatics*, 19(7):796–802, 2003.

[211] R. M. Wartell and A.S.Benight. Thermal denaturation of DNA molecules: A comparison of theory with experiment. *Phys. Reports*, 126(2):67–107, 1985.

[212] H. Werntges, G. Steger, D. Riesner, and H.-J. Fritz. Mismatches in DNA double strands: thermodynamic parameters and their correlation to repair efficiencies. *Nucleic Acids Res.*, 14(9):3773–3790, 1986.

[213] J. Wilhelm, A. Pingoud, and M. Hahn. Sofar: Software for fully automatic evaluation of real-time PCR data. *BioTechniques*, 34:324–332, 2003.

[214] J. Wilhelm, A. Pingoud, and M. Hahn. Validation of an algorithm for automatic quantification of nucelic acid copy numbers by real-time polymerase chain reaction. *Anal. Bioc.*, 317:218–225, 2003.

[215] D. M. Williams, S. Koduri, Z. Li, W. D. Hankins, and I. Poola. Primer design strategies for the targeted amplification of alternatively spliced molecules. *Anal. Biochem.*, 271(2):194–197, 1999.

[216] M.L. Wong and J.F. Medrano. Real-time PCR for mRNA quantitation. *Biotech.*, 39:75–85, 2005.

[217] P. Wu, S. i. Nakano, and N. Sugimoto. Temperature dependence of thermodynamic properties for DNA/DNA and RNA/DNA duplex formation. *Eur. J. Biochem.*, 269:2821–2830, 2002.

[218] P. Wu and N. Sugimoto. Transition characteristics and thermodynamic analysis of DNA duplex formation: a quantitative consideration for the extent of duplex association. *Nucleic Acids Res.*, 23(23):4762–4768, 2000.

[219] D. Xu, G. Li, L. Wu, J. Zhou, and Y. Xu. Primegens:robust and efficient design of gene-specific probes for microarray analysis. *Bioinformatics*, 18(11):1432–1437, 2002.

[220] W. Xu, W. J. Briggs, J. Padolina, W. Liu, C. R. Linder, and D. P. Miranker. Using MoBioS' scalable genome join to find conserved primer pair candidates between two genomes. *Bioinformatics*, 20:i355–i362, 2004.

[221] N. Ben Zakour, M. Gautier, R. Andonov, D. Lavenier, P. Veber M-F. Cochet, A. Sorokin, and Y. Le Loir. GenoFrag: software to design primers optimized for whole genome scanning by long-range PCR amplification. *Nucleic Acids Res.*, 32(1):17–24, 2004.

[222] Y. Zeng, A. Montrichok, and G. Zocchi. Bubble nucleation and cooperativity in DNA melting. *J. Mol. Biol.*, 339:67–75, 2004.

[223] A. Zien, G. Ratsch, S. Mika, B. Scholköpf, T. Lengauer, and K.-R. Muller. Engineering support vector machine kernels that recognize translation initiation sites. *Bioinformatics*, (9):799–807, 2000.

[224] H. Zipper, H. Brunner, J. Bernhagen, and F. Vitzthum. Investigations on dna intercalation and surface binding by SYBRGreen I, its structure determination and methodological implications. *Nucleic Acids Res.*, 32(12):e103, 2004.

APPENDIX A: FILTERS

We use four filters for measuring sequence similarity in binding site search. The simplest counts the number of mismatches between two k-mers, and the most complicated computes the binding energy of the reverse complement of a k-mer binding to the query according to the partition function model. The other two filters are described in the next two subsections. We use A and B to represent the sequences input to the filter; these sequences have length m and n, respectively. We use A_i to represent the ith element of sequence A.

Free Energy Filter

The free energy filter is defined first by mapping sequences A and B to complex valued vectors $\Phi(A)$ and $\Phi(B)$, and then taking their inner product. We developed the mapping Φ and present it here for the first time.

The mapping function has the property that if A and B are identical, then

$$\langle \Phi(A)^*, \Phi(B) \rangle = \Delta G(A, \hat{B}) + \Delta G(B, \hat{A}) - \Delta G_i$$

where a carat denotes reverse complement, and $\Delta G(A, \hat{B})$ is the free energy of binding of A to the reverse complement of B, and ΔG_i is a duplex initiation energy parameter. This computation of the binding energy between two sequences approximates the free energy computations presented in [174].

The inner product $\langle \Phi(A), \Phi(B) \rangle$ has the property that the angle between $\Phi(A)$

Table 7.1: Loop penalty matrix for Alignment Filter.

1.050	0.120	0.010
0.120	0.800	0.003
0.010	0.003	0.003

and $\Phi(B)$ increases with the number of mismatches. The angle is also sensitive to the identity of the mismatching bases, and will increase more for strongly destabilizing mismatches (such as C—C) than for mildly destabilizing mismatches (such as G—G).

The inner product can be computed as

$$\langle \Phi(A), \Phi(B) \rangle = \sum_{k=1}^{m-1} [\Delta G_s(A_k, A_{k+1}) \Delta G_s(B_k, B_{k+1})] \\ + [\delta(A_k = B_k)\delta(A_{k+1} = B_{k+1})]$$

where $\Delta G_s(A_k, A_{k+1})$ is the free energy of binding of the dinucleotide stack [174].

Alignment Filter

We designed the alignment filter to coarsely approximate the partition function model of DNA binding. This filter computes a score that rewards runs of consecutive identical bases in each sequence, and that penalizes loops analogously to the loop entropy functions in [174]. The parameters that we use to reward consecutive matches and penalize loops were optimized for this application.

The filter value is computed first by filling a dynamic programming matrix, and then computing the sum of all of its entries. This filter uses an AT reward parameter α, and a GC reward parameter β. We set $\alpha = 1.1$ and $\beta = 1.15$. This is analogous to assigning a slightly more stable energy to GC base pairs than AT base pairs, but this filter neglects specific dinucleotide effects.

The dynamic programming matrix is filled in as follows. If A_i is not equal to B_j, then $F_{i,j}$ is set to zero. Otherwise, if A_i is equal to B_j, then

$$F_{i,j} = \max_{i-3 \leq x < i, j-3 \leq y < j} (R * F_{x,y} * L[i-x, j-y])$$

where $R = \alpha$ if A_i and B_j are both A or T, and $R = \beta$ otherwise. The loop penalty matrix L is given in Table 7.1. The element in the first row and column is greater than 1 in order to reward consecutive matches.

VITA

Tobias Patrice Mann earned a Bachelor of Science in Computer Science and a double Bachelor of Arts in Mathematics and Philosophy in 1997, a Master of Science in Electrical Engineering in 2002, and a Ph.D. in Genome Sciences in 2007, all from the University of Washington.

VDM publishing house ltd.

Scientific Publishing House
offers
free of charge publication

of current academic research papers, Bachelor´s Theses, Master's Theses, Dissertations or Scientific Monographs

If you have written a thesis which satisfies high content as well as formal demands, and you are interested in a remunerated publication of your work, please send an e-mail with some initial information about yourself and your work to *info@vdm-publishing-house.com*.

Our editorial office will get in touch with you shortly.

VDM Publishing House Ltd.
Meldrum Court 17.
Beau Bassin
Mauritius
www.vdm-publishing-house.com

VDM Verlag Dr. Müller | LAP LAMBERT Academic Publishing | SVH Südwestdeutscher Verlag für Hochschulschriften